Urban Regimes and Strategies

University of Chicago Geography
Research Paper no. 239

Series Editors

Michael P. Conzen
Chauncy D. Harris
Neil Harris
Marvin W. Mikesell
Gerald D. Suttles

Titles published in the Geography Research Papers series prior to 1992 and still in print are now distributed by the University of Chicago Press. For a list of available titles, see the end of the book. The University of Chicago Press commenced publication of the Geography Research Papers series in 1992 with no. 233.

Urban Regimes and Strategies

Building Europe's
Central Executive District in Brussels

Alex G. Papadopoulos

The University of Chicago Press

Chicago and London

Alex G. Papadopoulos is assistant professor in the Department of Geography, DePaul University.

The University of Chicago Press, Chicago 60637
The University of Chicago Press, Ltd., London
© 1996 by The University of Chicago
All rights reserved. Published 1996
Printed in the United States of America
05 04 03 02 01 00 99 98 97 96   1 2 3 4 5
ISBN: 0-226-64559-2 (paper)

Library of Congress Cataloging-in-Publication Data

Papadopoulos, A. G.
    Urban regimes and strategies : building Europe's central executive district in Brussels / Alex G. Papadopoulos.
        p.   cm. — (University of Chicago geography research paper ; no. 239)
    An extensive revision of the author's thesis (Ph.D.), University of Chicago, 1993, with title: The Quartier Européen-Léopold, Brussels.
    Includes bibliographical references and index.
    1. City planning—Belgium—Brussels. 2. Central business districts—Belgium—Brussels. 3. European Economic Community.
    I. Title. II. Series.
    HT169.B42B796   1996
    307.3'3—dc20   96-12573
                                                                                CIP

To my teachers, with gratitude

Contents

Figures

Tables

Acknowledgments

This study of urban change and political-spatial brokering was conceived in the winter of 1989, in the bowels of the archive of the old Ministry of Colonies, in place Royale, Brussels. Full of apprehension about my projected doctoral fieldwork project in Zaire's Shaba Province, I marveled, first with half a mind, about the tremendous urban changes that were occurring in a city I had come to know well since I was a child. These changes extended from just outside the archive's doors to the quartier Léopold, and to some extent, throughout the Brussels agglomeration. In the end, my curiosity about the transformation of Brussels by European integration activities won out over my Central African research. This volume is based, in part, on a doctoral dissertation that explored some early questions about the dramatic morphological and functional transformations in Central-East Brussels. I owe many thanks to a great number of people who helped me research and write this book, but I want to extend the warmest of thanks to Sarah Peirce for encouraging me to plow through, and not to rest until it was done.

My greatest debt of gratitude is to my teachers at the University of Chicago: Professor Michael P. Conzen, who acted as dissertation director, research advisor, thoughtful critic, and teacher of urban morphology. My greatest intellectual debt is to him. I am indebted to Professors Marvin W. Mikesell and Chauncy D. Harris for carefully reading early drafts of this study and offering much constructive criticism throughout the period of research and writing. I am also grateful to Professor Norton S. Ginsburg for giving me much needed encouragement, and most importantly, for introducing me to human geography when I was his student in the Committee on International Relations. It is to all of them that I would like to dedicate this book.

During my stay in Brussels I received the assistance and counsel of many people whom I cannot possibly thank here individually. I would like to thank the librarians of the Royal Library Albert I, and especially Mme Lisette Danckaert for guiding me through the riches of the cartographic collection. I also want to thank the officials of the cadastral service of the Brussels Agglomeration, the National Institute of Statistics, the National Geographical Institute, the Commission of the European Union, the European Parliament, the Atelier de Recherche et d'Action Urbaine, Inter-Environnement Brussels, and a multitude of real-estate development firms, including Jones Lang Wootton and Richard Ellis, who provided me with the necessary information and commentary to complete this study.

Finally, I would like to extend a special thanks to my colleagues at the Department of Geography at DePaul University for their moral support, and especially to Professor Donald Dewey for being a nurturing chairman and a champion of this book project. Also, I would like to thank Ms. Carol Saller and Ms. Julee Tanner of the University of Chicago Press for expertly refashioning my manuscript into this book and putting up with me during the copyediting stage. Needless to say, all blunders and omissions in it are of my own doing.

Prologue

I was born in Frederikshavn, Denmark. I am 24 years old and graduated in political sciences two years ago. I'm a trainee with the General Secretariat of the European Commission. I share my apartment with a German student. My name is Ingeborg, and I am happy to live in Brussels.

Brussels and I, we go back a long way. I wasn't born here; Ramsgate, Kent is where I'm from. Trained as an agrochemical engineer, I was sent over as a specialist in 1973 to represent the United Kingdom in the making of Green Europe. My name is Graham and I'm happy to live in Brussels.

When I left Lisbon, I was supposed to replace a journalist colleague of mine for about three weeks. I've been a permanent correspondent at the European Communities for six years now. My name is Joao, and I'm happy to live in Brussels.

Europe, an idea which is being built in Brussels.

—Excerpts from a 1993 promotional brochure
of the Brussels-Capital Region

Unlike Ingeborg, Graham, and Joao, she does not want to have her name published, not because she shies away from public scrutiny or, for that matter, notoriety, but because her cause is not purely personal. She is the spokesperson for a group of inhabitants—most of them artists and academics—of the streets Hydraulique, Marie-Thérèse, de la Charité, and du Marteau, organizing against what they perceive as the wholesale and unwanted transformation of their neighborhood. The streets in question border the Brussels quartiers Léopold and Nord-Est, which have become a hotbed of urban transformation by international developers following the so-called "relaunching of Europe" during the first Delors presidency of the European Community.

Since 1985, these long-standing residents of the northwestern periphery of the growing European Community administrative park have observed the gradual absorption of the nineteenth-century city immediately to their south and southeast and its functional and aesthetic transformation. They are alarmed by the proliferation of office buildings in a part of Brussels renowned for the tranquillity and middle-class prosperity of its residential spaces. By January of 1992, office buildings—more precisely, the new facility of the Banco di Roma on the rue du Marteau—rose over the rowhouses of the protesters.

On 8 January of that same year, the protesters addressed their concerns in a letter to Guy Cudell and the Consultative Committee on Urban Planning of the municipality of Saint-Josse, responsible for granting plotting and building permits in their streets. This was one in a series of poignant albeit not very effective letters to their local representatives. In it they vented their anger about a permit allowing the construction of buildings poised to encircle the Ateliers Mommens—an artists' colony and neighborhood symbol. The fact that the permit was awarded to a real-estate development firm in the immediate aftermath of the 24 November 1991 municipal election added insult to injury. With their electoral capital spent until the following round of elections, the protesters have a limited arsenal against what they see as the condominium of big real-estate interests, Regional and municipal governments, tertiary sector firms, and the European Union executive branch resident in the quartier.

The protesters are concerned that traffic congestion, which is already a notable problem throughout the city, will increase with further urban development. It brings with it the specter of long bottlenecks, air and noise pollution, and deteriorating street pavements. The tranquil residential space created and enhanced by the convergence of rowhouse gardens in the interior of street blocks throughout the nineteenth-century extensions of the city will be compromised, as trees will suffer in the midst of pollution and tall buildings. The very physical fabric of their neighborhood, the old garden walls, the houses themselves, the protesters claim, will be compromised by vibrations from automobile traffic and building construction. Consequently, they demand assurances from the municipal authorities that their way of life is not going to change, and if it has to, that the impacts are going to be minimized.[1] The Ateliers Mommens have finally made their debut into the world of collapsing trade barriers, European union, and world-class real-estate development deals.

It is easy to decry the displacement of residents and the disfigurement of picturesque neighborhoods, especially when the antagonists are Brussels bohemians and impersonal corporations. What will be described here, however, is a contentious process of urban transformation that far transcends the concerns of the Ateliers Mommens, the particular real-estate developer, and the officers of the Consultative Committee on Urban Planning of Saint-Josse. The case accounted above is incidental to a general recasting of a portion of central Brussels on the basis of a new conception of the modern Western downtown—a specialized "central executive" as opposed to a more inclusive "central business" district. Fundamental to the understanding of this new urban form is the mechanism which has given rise to it: regimes or urban alliances evoking John Logan and Harvey Molotch's powerful model of growth coalitions and land-dealing entrepreneurs but including structures and motivations which were not considered by them.

The regimes which are transforming the quartier differ from the "public-private" partnerships which are managed by not-for-profit development corporations or parastatal organizations that are common in North American metropolises.[2] As described by Peter Michael Smith, these institutions organize an overt alliance between the local authorities and certain business constituencies with land-based interests under the rubric of partnership.[3] By contrast, the quartier's regimes appear at the very least informal in that they are covert and ephemeral, as they have been created to carry out discreet spatial tasks within a prescribed time frame. Unlike the sinister, and uncertainty- and risk-plagued partnerships described by Michael Smith and Susan and Norman Fainstein, which hope to attract firms to the city, the quartier's regimes leave little to chance, as the participation of initiators, strategists, contractors, and user-consumers of the

---

1. Letter by the inhabitants of the streets Hydraulique, Marie-Thérèse, de la Charité, and du Marteau addressed to Guy Cudell and the Commission de concertation de l'urbanisme de la Commune de Saint-Josse (8 January 1992).

2. Susan S. Fainstein, Norman I. Fainstein, Richard Child Hill, Dennis Judd, and Michael Peter Smith, *Restructuring the City: The Political Economy of Urban Development* (New York: Longman, 1983); Michael Peter Smith, *City, State, and Market: The Political Economy of Urban Society* (New York: Basil Blackwell, 1988), pp. 198–200.

3. Smith, p. 210.

project is a sine qua non for the project's launching. Whereas the typical urban coalition courting foreign and out-of-town businesses disburses tax abatements, property and sales-tax holidays and rollbacks, lease financing, and tax-exempt industrial revenue bonds, the quartier's regime players visualize, produce, consume, and manage for a profit projects designed for economic and political growth. This is a study of these regimes, the idiosyncratic executive business district they have given rise to, and the impact of European integration on urban form and function in millennial Brussels.

1
Introduction

E VERY business day during the daytime hours, European Union cadres, lobbyists, lawyers, management and tax consultants, venture capitalists, diplomats, government ministers, members of different parliaments, including the European Parliament, trading and industrial delegates, representatives of labor organizations, translators, interpreters, computer programmers, librarians, secretaries, telephone operators, security guards, and maintenance personnel, along with an assortment of store employees and managers, congregate in the Brussels quartiers Léopold and Nord-Est, and the area surrounding the Schuman roundabout, to administer, lobby, defend, examine, negotiate, represent, cajole, pressure, translate, interpret, program, catalogue, file, and maintain European Union policy, people, and buildings, as the case may be.

The quartier Léopold and the adjacent areas to the north, east, and northeast (most of the late-nineteenth-century quartier Nord-Est) contain the majority of the physical facilities of the executive branch of the European Union. This area of approximately eighty-five street blocks is situated directly to the east of the medieval, pentagon-shaped city and the Belgian government administrative park. Developed since the formation of the European Community in 1957 as the provisional seat of its executive branch, the newly transformed quartier has consumed a portion of the nineteenth-century extension of the city. While the quartier's street plan has remained largely unchanged since the entire extension was completed at the turn of the century, the presence of the European institutions and the increasingly visible international tertiary and quaternary sectors have changed the politics, economics, urban morphology, and aesthetics of the quartier.

I have chosen this area of Brussels, which for the contemporary period I have labeled "quartier Européen-Léopold," as the focus of a study of urban morphological change and the processes that have led to the emergence of a new type of executive central business district. The conclusions of this study suggest that these processes may have significant implications for the future development of the city of Brussels and other "world cities."

The installation of a major economic, administrative, and political institution in the midst of a predominantly residential area of a medium-sized Western European capital gives us the opportunity to witness the physical and functional transformation of both its immediate area and the city as a whole. Much like dropping dye into a clear pool of water, the insertion of a powerful pole of attraction and transformation such as the European Commission and its appended agencies acts as a major catalyst for economic, social, and political change that ripples throughout the quartier and the city. This institutional stimulus reacts with the peculiar planning and real-estate development environment of Brussels, as well as with the city's political power structure, to produce a distinctive case of central business district formation.

Whereas the nineteenth-century quartiers Léopold and Nord-Est were predominantly residential in character up to 1957, they have since been changing into nothing less than a significant alternative central business district in competition with the traditional district inside the ancient pentagonal city, the avenue Louise, the recently expanding quartier Nord, and a collection of suburban and extraurban office development sites. This nascent central business district unsettles its immediate area, granting access to a host of transforming agents: international institutions, Belgian and foreign banks and corporations, investors and speculators of all magnitudes, officials of the federal state, the Brussels-Capital region and the municipality of Brussels, urban planners, and quartier activists. The manner in which these agents interact with each other in regimes of cooperation or resistance controls the reorganization of both the transformed areas and the surviving nineteenth-century urban fabric of the quartier.

The evolving CBD introduces new functions, new pressures, new technologies, and, therefore, new relationships between agents of change and structures. An important new feature is the securitization and the globalization of the international real-estate market, which allows Brussels real estate to be marketed in international securities markets worldwide. The nascent "European" CBD introduces a new nomenclature of functions and physical aspects, and a new direction in the succession of functions and morphological changes in the inherited landscape of the quartier. Whereas the quartier Européen-

Léopold exhibits certain characteristics of the typical central business district (such as centrality and density, a great variety of high-level administrative functions, and prestigious and high value-added corporate activities), it lacks any notable commercial activity aside from a cross-section of small business-related commercial establishments such as messenger services, office supplies stores, travel agencies, and snack bars catering to the business lunch crowd. The rising costs of corporate establishment in Brussels, and particularly in the quartier Européen-Léopold, increasingly prohibit the introduction of non-specialized and "back-office" operations.

Historical Background

Starting modestly as the political and administrative capital of the duchy of Brabant at the end of the thirteenth century, Brussels attained a liberal civic culture that allowed it to flower economically and artistically as a center of Flemish culture. It retained a regional significance through the centuries of dynastic rivalry in Europe and emerged in 1830 as the national capital of an independent Kingdom of Belgium. In the contemporary Belgian political environment, Brussels continues to be the capital of a recently federated Belgium, and an important official symbol of Belgian national identity (fig. 1). In terms of contemporary international affairs, Brussels has added to its federal/national identity that of European integration and federalism. The understanding of CBD morphogenesis in the quartier Européen-Léopold is well served by an awareness of the political and institutional stimuli of European integration, because this is not a typical CBD, and certainly differs from the CBDs one encounters in other European or American cities.

Despite the general rejection of European federalism as utopian thinking, the continuity and resilience of the federal idea in the political and constitutional development of the Union have been inescapable. While most recent Anglo-American political science literature on Community affairs depicts it as little more than an intergovernmental grouping of independent states,[1] the Community has

---

1. H. Wallace, W. Wallace, and C. Webb, eds., *Policy Making in the European Community* (London: John Wiley, 1983); M. Forsyth, *Unions of States: The Theory and Practice of Confederation* (Leicester: Leicester University Press, 1981); P. Taylor, *The Limits of European Integration* (London: Croom Helm, 1983); and C. Tugendhat, *Making Sense of Europe* (London: Viking Penguin, 1986).

been rooted in a traceable and significant political ideology and strategy which has been the foundation of the post-Gaullist direction it has taken in the last twenty years.[2] Intergovernmental negotiations between national political bodies are indeed the backbone of the operations side of the European Community and take place from the administrative to the ministerial level in Brussels, Luxembourg, and Strasbourg. That fact does not, however, negate the dual nature of the European construction. In the words of Christopher Tugendhat, member of the European Commission (1977–85):

> Federalism and intergovernmentalism, supranationalism and cooperation between different nationalities: two concepts of Europe . . . have been vying with each other since the earliest days of the Community . . . [T]he founding fathers allowed neither to prevail in the Treaties of Rome, which represents a delicate balance between the two.[3]

With the Single European Market now a fait accompli, we need to take into account the reality of the unfolding Regional reorganization of European space-economy and polity. Excursions into monetary and fiscal policy (Fouchet Plan: rejected 1961), foreign policy (Werner Plan: rejected 1970; Tindemans Plan: rejected 1976), and the harmonization of social and Regional development programs, among others, appeared unacceptable a few years ago and were deemed contrary to the sanctity of the nation-state by statesmen such as Charles de Gaulle and Margaret Thatcher. Today they are becoming binding policy throughout the European Union. The nature of the current executive agenda, including the efforts of the Union to speak in one voice about such matters of international significance as the boycotting of Iraq in the aftermath of the Persian Gulf War, the breakup of Yugoslavia, and the economic restructuring of Russia, indicates

---

2. The early idea of European federalism and the intellectual elite which espoused it appear to have emerged during the Second World War. In the famous Ventotene Manifesto of 1941, Altiero Spinelli made the first declaration devoted to European Unification. Later he would point out that the consensus in the resistance movement was that it would be preferable to give a federal structure to Europe since this would also solve the problem of coexistence in peace and freedom with Germany. A. Spinelli, "European Union in the Resistance," *Government and Opposition* 2 (1966–67): 325. The federalist view is presented in great detail by M. Burgess, *Federalism and European Union: Political Ideas, Influences, and Strategies in the European Community, 1972–1987* (London: Routledge, 1989); M. Burgess, ed., *Federalism and Federation in Western Europe* (London: Croom Helm, 1986).

3. C. Tugendhat, *Making Sense of Europe* (London: Viking Penguin, 1986), p. 71.

Fig. 1. Belgium following federalization

that the Union is slowly moving to fulfill the federalist obligation implicit in the Treaties of Rome. If, therefore, the "1992" program for the creation of a Single European Market was a triumph for functionalists like Jean Monnet, the Maastricht Treaty is a vindication of early federalists like Altiero Spinelli (fig. 2).

As important as these developments may be for Europe as a whole, European integration and possible federalization have a special significance for the city of Brussels. Regardless of whether "Europe 1992" suggests in political science jargon "intergovernmental interaction between political élites," or the first phase in the construc-

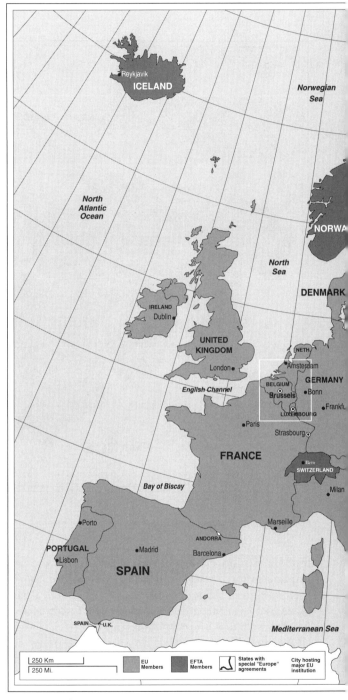

Fig. 2. General orientation map

tion of a federal European union, the implications for the Brussels urban landscape and socioeconomic character are significant.

Changes in the quartier Européen-Léopold have taken place with such speed that a mere twenty-eight years after the construction of the first building in the quartier, the urban canvas is practically filled with new urban forms. While several urban plans have been recommended by the Regional government, urban developers, and pressure groups, the quartier is being shaped by an urban coalition dominated by private-sector interests. It is often difficult to discern the impact of European federalist ideology on the new morphological character of the quartier. It has been equally difficult, at least at this stage, to relate a continuum of morphological change to a continuum of European integration.[4] It is instead the intensity and nature of activity on the part of the urban coalition that can be seen as related to morphological change. The interaction between the new European ideology and the ideology of the quartier's urban coalition raises questions about the future shape of the quartier that for the time remain unanswered.

The City in Its National and Regional Urban Geographic Context

The evolution of a central executive district in the quartiers Léopold and Nord-Est is taking place within a more general framework of social and urban geographic reorganization of the Brussels agglomeration and the significantly grander area that makes up its urban field. Moreover, one could now fairly safely associate functional changes in Brussels and other major European Union urban centers with European integration regional development, transportation, and telecommunications policies.

---

4. The likelihood that Brussels will attract all or the majority of European institutions in the future—always assuming that European integration proceeds smoothly—is small. Member states are fully aware of the economic and political dividends in prestige that the placement of such institutions produces and will most certainly jockey to attract them. The movement of all committee and subcommittee meetings of the European Parliament from Strasbourg to Brussels thanks to the influence of the Commission suggests a centralizing trend. Some of the new planned institutions such as the much talked-about "Euro-Fed" (European Central Bank) will nonetheless have enough political mass or influence to resist the pull of Brussels. In all probability, the European Central Bank will be founded in Berlin or Frankfurt, Germany, in recognition of that country's leading role in the Community.

The aggressive promotion of the high-technology "European net-works," which are expected to bring the European partners into an ever closer functional union, suggests that the Union leadership has both the vision and the intent to make major European cities the cor-tex of the new unified market. The flourishing of transnational super-train networks, such as the one spearheaded by the Société Nationale des Chemins de Fer (SNCF) of France—the Train à Grande Vitesse or TGV—illustrates the point. Originally a French scheme featuring a rapid link between Paris and Marseilles, by the summer of 1995, the TGV will be carrying passengers throughout a growing network of lines extending from the Netherlands to Spain. The Brussels link to London will make use of the Euro-Tunnel under the English Channel and will slash travel time to Paris by 50 percent. With time-distance between such key urban centers drastically reduced, the delimitation of Brussels's urban field will soon have to be reassessed.

At the national level, shifts in manufacturing investment from Wallonia to Flanders and specifically in the triangular area described by Brussels, Antwerp, and Ghent, have been responsible since 1960 for population mobility out of and around the capital. Within this key industrial zone, the axial belt connecting the capital with Antwerp, a world-class port rivaling Rotterdam in capacity, has been the site of greatest foreign industrial investment in Belgium during the eco-nomic expansion of 1984–90.

The structural influences at the European Union level notwith-standing, it is important to place the seemingly localized phe-nomenon of the quartier Européen-Léopold within a regional context of urban geographic change. The changes in the city since the end of the Second World War are repeated in many other Western Euro-pean metropolises and resemble to some degree changes that have occurred in North American cities.

What clearly differentiates Brussels and other like European metropolises from their North American counterparts is the existence of compact, old, and sometime ancient cores that influence the devel-opment of central functions in the modern era. In the case of Brussels, the densely built-up area was formed during a number of historical eras of construction which largely conclude with the Second World War. Following the growth of urbanization rates during the nine-teenth century and the physical expansion of cities and their centers, Brussels entered the suburbanization phase earlier than most other Western European metropolises, partly owing to the existence of an immensely dense and efficient transportation network (fig. 3).

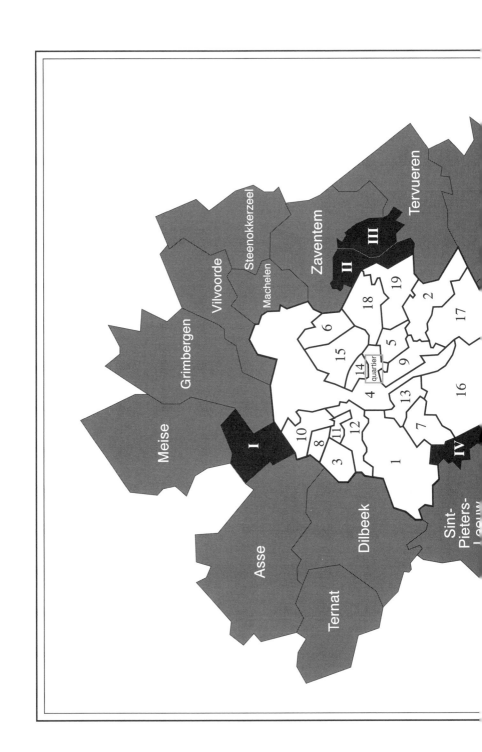

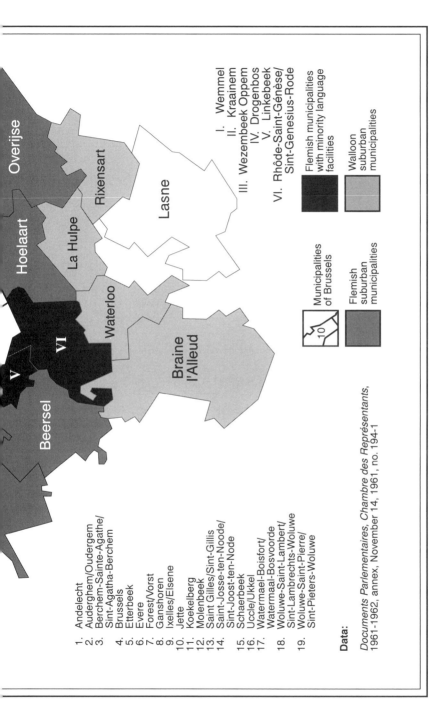

1. Andelecht
2. Auderghem/Oudergem
3. Berchem-Sainte-Agathe/
   Sint-Agatha-Berchem
4. Brussels
5. Etterbeek
6. Evere
7. Forest/Vorst
8. Ganshoren
9. Ixelles/Elsene
10. Jette
11. Koekelberg
12. Molenbeek
13. Saint Gilles/Sint-Gillis
14. Saint-Joost-ten-Noode/
    Sint-Joost-ten-Node
15. Schaerbeek
16. Uccle/Ukkel
17. Watermael-Boisfort/
    Watermaal-Bosvoorde
18. Woluwe-Saint-Lambert/
    Sint-Lambrechts-Woluwe
19. Woluwe-Saint-Pierre/
    Sint-Pieters-Woluwe

I.   Wemmel
II. Kraainem
III. Wezembeek Oppem
IV. Drogenbos
V.   Linkebeek
VI. Rhôde-Saint-Génèse/
     Sint-Genesius-Rode

Municipalities
of Brussels

Flemish suburban
municipalities

Flemish municipalities
with minority language
facilities

Walloon
suburban
municipalities

Data:

Documents Parlementaires, Chambre des Représentants,
1961-1962, annex, November 14, 1961, no. 194-1

Fig. 3. The Brussels-Capital Region and surrounding municipalities

Herman Van der Haegen accounts very well the characteristics of and the factors influencing the demographic decline of the central city. He describes it as the result of declining fertility among the Belgian Bruxellois, the enhanced financial picture of Belgians in general, the desire for a green living environment, the deterioration of the nineteenth-century housing stock, the worsening traffic congestion, the loss of industry and employment through national and world economic restructuring, and the population's disaffection with the city's political administrative framework.[5] Van der Haegen's account is not critical of the changes. While he aims for impartiality, he reveals his preferences by stating that the Belgian urban evolution process has reached a very advanced stage, and according to van der Berg et al., "the most advanced in Europe."[6]

A decade after the end of the Second World War, Brussels was rapidly losing population, mainly to the attractive green municipalities in its immediate periphery but also to the urban fringe. The demographic growth of the Brussels metropolitan area (the nineteen municipalities that make up the Brussels-Capital Region) peaked in 1970 and since then the Region as a whole has been losing population to more distant municipalities, with the central city experiencing the most severe loss.[7] Partly responsible for this decline was the erosion of the industrial base of the Capital Region. Raymond Riley cites statistics of the Belgian Ministry of Economic Affairs and Energy illustrating the marked shift in the capital from a blue-collar to a white-collar economic structure: "Between 1970 and 1972 . . . of the fifty-eight new foreign industrial investments in Brussels, no [fewer] than forty-seven were merely offices of plants established outside the Region"[8]

This process of suburbanization and even deurbanization explains

---

5. Herman van der Haegen, "The Crisis of the Inner Cities in Belgium," in Günter Heinritz and Elisabeth Lichtenberger, eds., *The Take-Off of Suburbia and the Crisis of the Central City: Proceedings of the International Symposium in Munich and Vienna 1984* (Stuttgart: Franz Steiner Verlag Wiesbaden Gmbh, Erdkundliches Wissen, vol. 76, 1986), pp. 196–97.

6. Ibid., p. 194; Leo van der Berg et al., *Urban Europe: A Study of Growth and Decline* (Oxford: Pergamon, 1982).

7. P. Cabus, "De stedelijke ontwikkeling in Vlaanderen, 1947–1976: Vaststellingen en indicaties voor een stedelijk beleid," *GERV-berichten* 29 (1980): 109–51.

8. Ministère des Affaires Economiques et de l'Energie, *Rapport 1970, 1971, 1972* (Brussels: 1971–73), in Raymond Riley, *Belgium* (Folkstone, UK: Dawson, Studies in Industrial Geography, 1976), pp. 115–16.

radical shifts in the social economic geography of Brussels. Walter De Lannoy and M. Rampelbergh have mapped the effects on the central city in their *Sociaal-geografische atlas van Brussel en zijn randgebied*.[9] In brief, the center is being abandoned by the affluent, the young, the professionals, the native-born Belgians, the owner-occupiers of houses. The central city is becoming increasingly more "gray," owing to the great numbers of elderly, single women remaining as long-term occupants of houses; poorer, owing to the influx of working-class immigrants; and ethnically more complex, mainly attracting new North African residents. By contrast, the periphery—especially the eastern and southeastern municipalities—is the domain of prosperous Belgians and Western foreigners. Moreover, that same periphery is becoming rapidly a favored location for the tertiary and quaternary industry.

The end result is that the quartier Européen-Léopold phenomenon as a new-type regeneration of the central city—or at least a portion of it—flies in the face of this seemingly measured abandonment of the city center. For the time being, it appears that the urban transformation of the quartier works much in terms of Fainstein and Fainstein's "conversions" of declining waterfront industrial centers, such as Boston and Baltimore, and in the following manner: long-standing residential communities have given way to new elite engendered uses of urban space. If the northward movement of investment and transformation does not stop at the rue du Marteau, the working-class population of mostly North African extraction, along with a small number of Belgian Bruxellois bohemians, will be displaced. As Michael Peter Smith points out, neighborhoods occupied by new immigrants, who often have migrated illegally, constitute a politically weak constituency. Arguably, they will yield when an important institutional actor bids for their space.[10] Although it is not necessary to adopt wholesale this "proletarians under siege" worldview, it is important to identify such displacements as negative externalities of the process of regime-managed urban change.

Given the short time that the quartier has functioned as a CED, it is unclear what its impact will be in the demographic structure of central and east-central Brussels. It would be important to see whether it can act as a catalyst for the resettlement of the environing

---

9. Walter De Lannoy and M. Rampelbergh, *Sociaal-geografische atlas van Brussel en zijn randgebied* (Brussels: Vrij Universiteit Brussel, 1981).

10. Smith, pp. 200–201.

neighborhoods by professionals or, in fact, provide a further stimulus for the wholesale abandonment of the area by long-time residents. The long-awaited 1990 census suggested the trend, but the following one will surely give conclusive evidence.

Study Parameters

While speculation about the future of the city has been a part of this study, my major preoccupation has been with the recent and ongoing transformations in the quartier Européen-Léopold and to a lesser extent with selected developments throughout the Brussels-Capital Region which influence the growth of the quartier. Inevitably, I needed to understand the history that gave rise to a succession of urban patterns, some enduring today, which have influenced the evolutionary course of the city. The issue of greatest interest to geographers and urban planners is whether the quartier indeed represents a new type of central business district, one that, as noted above, includes a number of the descriptors of the typical CBD and a combination of functions that are atypical of traditional CBDs. The elements absent from this CBD's nomenclature of functions and forms are as important as those that are present.

The stimulus for the study has been the changing aspect of the city from national to international capital, and the implications of this transformation for the spatial organization of the quartier. The context is the emerging post-Fordist[11] and interdependent world economy, the assertion of Europe as a major moderator of international relations, and the changing functional and morphological character of Western metropolises.

The quartier Européen-Léopold encompasses approximately eighty-five street blocks in the nineteenth-century quartiers Léopold and Nord-Est. Approximately 90 percent of this area falls under the jurisdiction of the municipality of Brussels, while the remaining 10

---

11. Knox and Agnew define Fordism as "the socio-economic system that links mass production with mass consumption[:] a tense but durable relationship between big business, big labor and big government [that] enabled Fordism to provide the basis for the long postwar boom and unprecedented rise in living standards throughout much of the capitalist world." Paul Knox and John Agnew, *The Geography of the World Economy* (London: Edward Arnold, 1989), p. 340. In consequence, post-Fordism describes the abolition or negation of this relationship and the onset of an era where firms adopt spatially flexible modes of production, which are often incompatible with the social contract notions of Fordism.

percent is divided between the municipalities to the south, Etterbeek and Ixelles. I have drawn the boundaries of this area arbitrarily but with care to include the totality of the affected area and its unaltered fringe (fig. 4). Most parties involved in the quartier are interested in firmly defining the limits of the European Union administrative park, or simply the maximum geographical extent of EU and EU-related physical facilities. The motive behind the clamor for a more precise delimitation of the park differs according to the source of information: the federal government pursues what it considers best for the country as a whole; the regional government wants to serve the interests of its constituency by controlling, though not necessarily limiting, the extent of "Europe in Brussels," that is, the presence of the European Union institutions; pressure groups and urban foundations concerned with "mixicity" of functions in the quartier—the provision of housing and the rigorous conservation of the nineteenth-century city—want to see an end to the areal expansion of the EC administrative park, which infringes upon the surrounding residential and residential/commercial quartiers. The private sector, on the other hand, and in particular the large British real-estate developers and international financial interests such as the Banque Indo-Suez/ Banque Générale and Citibank, want the global real-estate market to shape and periodically reshape the boundaries of the European administrative park and the quartier surrounding it.

The veritable tug of war between bulldozing on one hand and restoration and renovation on the other, sanctioned by a peculiar Brussels urban planning tradition, transforms the quartier into a superspecialized central building district one lot or street block at a time. This planning style, which the more cynical will call nonexistent, makes the city a palimpsest for developers who with surgical precision can partially or completely erase the surviving built fabric to make room for new structures. I will show that this planning tradition is very much a part of the character of Brussels since its reinvention as a national capital.

To understand the urban morphological change of the quartier Européen-Léopold requires a careful review of the historical processes that created the nineteenth-century morphological frame and to some extent the ancient fourteenth-century pentagon-shaped city. Since the morphological processes do not always remove the total older urban fabric in order to replace it with a new one, we need to account for the role and influence of the remaining urban fabric in shaping the modern and radically different socioeconomic, political, cultural, and aesthetic environment of the quartier in the 1990s.

A detailed accounting of the urban landscape reveals the shape and age of the street plan, the various building types, their ages, amenities, values, in some cases their ownership, and, importantly, their land-uses. This tapestry of information shows that buildings reflect a variety of functions. From the economic perspective, they provide for investment, they store capital, create work, and house administrative and political activities. Socially, they appear to support relationships, provide shelter, express social divisions, and allow for the congregation of like-minded agents, permit hierarchies, house institutions, express status and authority, and embody property relations. Spatially, they establish place, define distance, enclose space, differentiate area and help redefine their surroundings. Culturally, they store sentiment, symbolize meaning, express identity, and embody taste. Politically, they symbolize power, represent authority or the plurality of opinion, and become an arena for conflict or compromise.[12]

Since the measurement of urban patterns is only a stepping stone to understanding their meaning, it was necessary to go beyond an elaborate classification of the contemporary built environment in the quartier and focus on issues that are particular to Brussels and that raise interesting questions about the interaction between the built environment and the structures and agents of change.

At the next level of analysis, I identified four interdependent processes behind succession in land use, building types, and management control.

- A planning tradition that sanctions the idiosyncratic development of the quartier
- A land market that is equally firmly linked to the international capital market and to local business interests
- A political process that gives rise to ephemeral local and international regimes of cooperation between political and economic elites
- An aesthetic vision and vocabulary dominated by considerations of profit making, the commodification of architecture, and the rejection of Western traditions of urban monumentality

These four processes are discussed separately in chapters 4 to 7 and are synthesized and criticized in the final chapter. The quartier is

---

12. List adapted from Anthony D. King, *Global Cities: Post-Imperialism and the Internationalization of London* (London: Routledge, 1990), p. 11.

presumed to be an example and model, representing a new aspect of Western European urbanism, that may illuminate change, processes, and agency in the late twentieth century.[13]

Methodology and Data: The study includes micro and macro levels of investigation. As noted above, a detailed exploration of the urban morphology of the contemporary quartier Européen-Léopold is the cornerstone of the microlevel investigative methodology. Available data for the period 1957–92 have been compiled on the following morphological descriptors:

I.   Age of the street plan and individual buildings
      Building types
      Amenities
      Architectural styles
      Architectural preservation registration
II.   Land-use at the lot level
      Zoning
      Value by square meter
      Initiating agent
      Transforming agent (developer, architecture and/or
         construction firm)

Group I descriptors deal with the material aspect of the quartier. Group II descriptors deal with agency and the economic and social character of the quartier.

Contemporary data are fairly complete. Data on the age of the street plan and of individual buildings were obtained from cadastral maps of the city's Department of Public Works. Information on amenities was available through the Institut National de Statistique for each census. Additional information on city-subsidized renovations and amenities was obtained through city services awarding the subsidies.[14] A classification of architectural styles was compiled by

---

13. Adapted from Dieterich Denecke, "Research in German Urban Historical Geography," in Dieterich Denecke and Gareth Shaw, eds., *Urban Historical Geography: Recent Progress in Britain and Germany* (Cambridge: Cambridge University Press, 1988), p. 29. Denecke discusses the analytic functional approach which led to detailed studies of urban fragments, and their morphogenetic and functional change, especially during the nineteenth and twentieth centuries.

14. In an effort to promote urban renewal in the different municipalities and to increase occupancy through an improvement in the quality of housing stock, the city provides generous construction subsidies to owner-occupiers and small investors interested in renovating and modernizing old housing. The subsidies or "prims" may

street-level observation. "Classés" buildings, or buildings considered part of the architectural heritage of Belgium, are listed in the Royal Registry for the Preservation of the Artistic Patrimony available through the Foundation Roi Baudouin. The alternative listing "Emergency Inventory of Historic Structures" is available through the St. Luc Archives of the St. Luc School of Architecture.

Excellent information was obtained from air photographs. The Institut Géographique National has periodically photographed all urban areas in Belgium at a scale of 1:2,500. I have procured for this study a set of air photographs of the quartier for the years 1950, 1956, 1969, 1978, 1988, and 1991. In the absence of air photographs dating from before 1950, I relied on historical maps, including maps of the Department of Public Works at a scale of 1:5,000, which depict the outlines of buildings.

Land-use information cannot be obtained through insurance and fire department maps, as Belgium does not follow the Anglo-American tradition of making such information public.[15] Complete land-use information and some occupancy information were obtained through field observation ("doorbell research"), census materials, chamber of commerce indices, and trade association listings. Changes in zoning classifications, which are usually announced as royal decrees, were obtained in both database and map form.

A cautionary note must be included concerning available information on zoning. While in theory zones appear firm on paper, in practice they are not. Violations of the zoning code are frequent, especially in high-growth areas like the quartier Européen-Léopold. Land-use data compiled through direct observation are consequently the most reliable source.

The most common type of violation is the clandestine conversion and letting or selling of properties designated for residential use as offices or commercial spaces. This practice may go unnoticed by the authorities for a period of months or years and not be reflected in

---

account for 30 percent of the costs of renovation and may be applied to the maintenance and beautification of street façades, insulation and replacement of roofs, installation of sanitary facilities, kitchens, and central heating. The deal the city offers investors is less attractive than it appears. Owner-investors benefiting from city subsidies of this sort are often then required to let part or the whole of the property to low-income families or low-income elderly for a set period of years.

15. In Belgium, real property insurance is the responsibility of both the property owner and the renter/occupier. Rates differ according to condition of the building, land-use, and level of hazard. The information is undoubtedly available in databank form but remains to date inaccessible to the general public.

public documents. Once apprehended, property owners have either to conform to zoning regulations or appeal in court. In the appeal process for the reclassification of individual properties we see the expression of conflict between interest groups: the profit-minded investor against the electorate-minding civic leadership.

One additional obstacle concerns the procurement of information on property ownership, as for the time being this remains confidential in Belgium.[16] Some sample data on property ownership in the area were procured nonetheless through real-estate agencies and direct observation. Buildings owned by large institutional investors, such as insurance companies, occasionally featured signs revealing the identity of the owner corporation. Additionally, aggregate data on ownership and property values can be found in secondary sources, such as the trade journal *L'Evenement Immobilièr;* in the most widely read financial daily *L'Echo de la Bourse;* and in the periodic sectoral assessments by Belgian banks, national mortgage granting institutions, and major real-estate development firms.

Aggregate estimates on real property values are readily available. Unfortunately, price information at the individual property level is difficult to procure for the following reasons: first, the published price is usually the price demanded by the owner. The speculative character of the market suggests that these prices may be 20 to 30 percent higher than the real market value. Second, it is not uncommon in the sale of small or medium-sized properties for the contractants to report a lower sale price on relevant contracts since this price serves as the basis on which the considerable public tax is assessed (12 to 15 percent of sale value depending on the municipality). The most useful value assessments are those established by reputable real-estate developers such as Jones Lang Wootton, although they are released only in aggregate form. I circumvented the problem arising from the lack of lot-level land prices by soliciting the opinions of urban development firms on this matter. Their responses have allowed me to compile maps of differential land value for the entire quartier.

Names and addresses of agents engaged in any type of urban transformation can be obtained in building plan applications filed with the authorities of the municipality. These applications list the date of plan submission, the site of proposed change, the description of the change and the architect's drawings, the decision by the local

---

16. This information is commonly available to notary publics and attorneys who prepare contracts for the sale and purchase of properties. The general public does not have access to this information, unless the owner makes some sort of public declaration.

Fig. 4. The quartier Européen-Léopold

EU office installations

Point of orientation for photographs

square Ambiorix

rue La Corrège

avenue Michel-Ange

rue Franklin

av. de la Renaissance

rue Charles Martel

rue St-Quentin

rue Stévin

boulevard Charlemagne

Tacluime

Charlemagne

Berlaymont

rue Archimède

avenue de Cortenberg

Triangle

avenue de la Joyeuse Entrée

Parc du Cinquantenaire

Rond-Point Schuman

Council of Ministers

rue Froissart

rue Breidel

avenue d'Auderghem

Breydel

av. des Nerviens

rue van Maerlant

Parliament

Park Léopold

Borchette

Borchette

rue de la Tourelle

rue J.A. Demot

rue du Cornet

rue Général Leman

authorities and the date of the decision, the names and addresses of most of the agents involved (initiator-applicant, architect, builder), and miscellaneous technical data including the extent of floor space, land use, and cost of the work.[17] Regrettably, one can access only those applications that are pending and are therefore required by law to remain part of the public record. Fortunately, information on "megaprojects" currently under construction, of fundamental importance to the study, was available.

Knowledge about the initiators of changes in the building fabric is as important as knowledge of property ownership. Foreign initiation or execution of an urban project differs in character from a change sponsored and executed by a local agent. We can generalize from Mike Freeman's study of initiators, architects, and builders in Wembley, Watford, Aylesbury, and Northhampton (each municipality located at an increasing distance from the city of London), which found that external firms have a lesser sense of place and tend to construct buildings different from or discordant with the local environment,[18] to suggest that the origins of the initiator of change in the Brussels quartier Européen-Léopold is an important descriptor of morphological change, especially since these initiators increasingly are foreign corporations.

A series of maps of the quartier depicting phases of building construction, land value, and land use will provide the material basis necessary to the discussion of the morphological evolution of the quartier.[19]

---

17. This is the listing of information appearing in the model building plan application cited in Mike Freeman, "Commercial Building Development: The Agents of Change," in T. R. Slater, ed., *The Built Form of Western Cities: Essays for M. R. G. Conzen on the Occasion of His Eightieth Birthday* (Leicester: Leicester University Press, 1990), p. 254. This type of building plan application conforms perfectly with its Belgian counterpart. The usefulness of building plan applications as sources of important morphological information has also been noted by P. J. Aspinal and J. W. R. Whitehand in "Building Plans: A Major Source for Urban Studies," *Area* 12 (1980): 199–203; and R. G. Rodger, "Sources and Methods of Urban Studies: The Contribution of Building Records," *Area* 13 (1981): 315–21.

18. Freeman, pp. 256–73.

19. Land-use in the first floor (rez-de-chaussée) occasionally differed from land-use in the second and upper floors. In fact, the apportioning of buildings into a commercial first floor and residential upper floors is very common, especially in traditional townhouses.

The Chapters

The study examines in chapter 2 the major theoretical paradigms that contribute to the formulation of the research question and the methodology. Sketches of land-rent theory and urban morphology provide the theoretical framework for the understanding of the transformation of the quartier from a residential space into a superspecialized business district. The discussion of each paradigm is tied to relevant aspects in the development of the quartier, which are then elaborated in subsequent chapters.

Chapter 3 traces the evolution of Brussels through three morphological periods: a Flemish/dynastic period, during which a substantial part of the morphologic frame is defined; a "French"/nationalist period, starting with Belgian independence in 1830, during which Brussels sheds much of its Flemish social and urban morphology in favor of an industrial and institutional functionalism represented by new urban forms in French taste; and finally an international/supranational period, during which the national institutional functionalism of the nineteenth-century is reinforced but also revolutionized by the city's new role as the seat of the European Union executive branch.

Chapter 4 discusses the historical framework of land market regulation in Brussels, which in theory, at least, prescribes the limits of action by urban actors. Brussels is an atypical Western European city because of the historical incidents that made it a capital city, because of its peculiar urban regulatory tradition—or lack thereof—and because of its recurrent flirtation with American-style urban development practices and planning.

Chapter 5 investigates the impact of the land market on the physical and functional form of the quartier Européen-Léopold. It first reviews the fundamental concepts that drive land markets in free market economies in general and the Brussels land market in particular. Second, it considers the different kinds of market participants by discussing the strengths and weaknesses of Logan and Molotch's framework on agency and market competition in Western cities. Third, it describes and discusses the land market of the quartier by considering its relationship to other markets abroad and to submarkets in Brussels, and its land-use morphology. Fourth, it presents and discusses the findings of an opinion survey, conducted by the author, of real-estate development firms and independent real-estate agents on the land value profile, the direction of physical expansion of the business/administrative sector of the quartier, and the structure of market demand.

Chapter 6 presents the agents and layers of political interaction that affect urban planning and development in Brussels and the quartier. While some of these critical political forces originate in the international and national political arenas, since the constitutional reform of the state in 1989, it appears that Regional, cultural community, and local interests and politics are expected to take on primary importance. These forces have produced political structures by which urban policy is decided and carried out. They include the national government, the Regional government as an element of the 1989 federal reorganization, the corporate linkages between major financial players in the Brussels market, and the real-estate market itself. They are articulated by key politicians into an urban coalition supporting the continuing growth of the specialized CBD activities in the quartier Européen-Léopold.

Chapter 7 explores the interactions among agents and structures through the device of rational-choice games. Markets, national interest, and corporate strategy ultimately become both context and ammunition for agents—real-estate developers, activists, local politicians, homeowners—who vie for control of the processes that shape the material and functional form of the quartier.

Chapter 8 first investigates the "shape" or urban morphological frame of the quartier and its power to further influence the processes of "shaping." Second, it discusses the relevance and significance of monumentality in the case of this particular urban European enterprise. Finally, it critiques the quartier's aesthetic impoverishment.

Chapter 9 synthesizes the key themes of the study, and provides a comprehensive listing of the findings. The four processes investigated in the study—the planning, land market, political, and aesthetic dimensions—are discussed in terms of the fundamental issue of tension between agency and structure. *Agency,* in the guise of key political figures and business persons, shaping economic choice and political decisions influencing the form and function of the Brussels CED, and *structure,* in the guise of planning regulations, shifting global market forces, the agenda of the European bureaucracy, and treaty environment, are combined in a working framework which recommends that *regimes of cooperation* are the key device by which individuals manipulate structures for their own benefit or to some specific purpose. The goal of the regimes described in this study is, of course, the sustained growth and evolution of the superspecialized CBD into a "central executive district."

2

Theoretical Foundations

T HE OBJECTIVE of this study is to discover how and why locational and planning choices being made at a certain place with a certain built legacy have an impact on spatial patterns and morphology. That place is a nineteenth-century residential quartier of Brussels which has been jerked into the vanguard of urban development and planning by the decision of the European Communities, at the invitation of the Belgian government, to install there the European Communities executive branch.

The quartier Européen-Léopold constitutes a land market that has been improved over the last one hundred and thirty years with the construction of a great number of residential structures and institutions such as hospitals, schools, and religious establishments. Its rent value has fluctuated over time but has presently reached unprecedented highs owing to its increasing centrality. The quartier is also a community of people that has fluctuated in size, changed in composition and occupation, and organized itself in a variety of ways to achieve certain goals. Finally, the quartier is a mosaic of physical structures, some dating to the middle of the nineteenth century, and an increasing number dating from the latest wave of expansion. These structures embody today's and yesterday's wealth, investment potential, status, corporate competition, international cooperation, social relationships, ethnic identity, and aesthetic values. This investigation addresses the evolution of the quartier from the green fields and *extra muros* ecclesiastical land of the early nineteenth century to a continental, and in some respects global, business district with very specific referents to the process of Regional economic and political integration taking place in the 1990s.

The research paradigms that are most useful to understanding the transformation of the quartier are land-rent theory and urban morphology. Land-rent theory demonstrates how we should look at the quartier as a land market, which is a view of great significance for a number of actors in the quartier: the real-estate developer seeking maximum rents, the corporate client jockeying for maximum central- ity, the European institutions shopping for favors, the national and Regional governments balancing the city's burdened budget with tangible and intangible returns from the EC presence in the quartier. At a much grander scale, the local land market is linked to the inter- national capital market and the now securitized global real-estate market. Urban morphology ties the consideration of the physical as- pect of the city (the ground plan, building forms, and land uses) with the processes and agents that have shaped and reshaped it. Urban morphology instructs that the city as a built environment is not ex- clusively a dependent variable, but rather an important interactive participant in landscape transformation along with the tendencies, in- trusions, and successions of economic, political, and social actors.

Both these theoretical traditions are intimately linked with eco- nomic and geographic approaches to urban transformation. Methodo- logically, however, the study of the quartier Européen-Léopold expands on the application of fundamentals of these approaches by integrating notions of rational choice-based decision-making and regime theories. This dialectic linchpin is explored largely empiri- cally in detail in the substantive chapter on political process in the quartier, since very little has in fact been written by rational-choice theoreticians that relates to urban management.[1] This exploration is not an extension of theory but a new application—this one not neces- sarily creating new game-theoretic rules, but revealing new impacts, spatial ones, which had either not interested or been ignored by rational-choice theoreticians. It is suggested here then, that the appli- cation of regime theory in an urban management context allows the integration of interactions between agents and structures in a manner traditional land-rent and urban morphology theories do not.

Land and Land-Rent theory

"Land," as has been defined by economists, encompasses at least two

---

1. Regime and rational-choice theories have explored decision making and the me- chanics of cooperation in statecraft, international politics and economics, institutional bargaining (for example, in the U.S. Congress), and the behaviors of optimizing con- sumers. Jon Elster has made exceptional strides in shedding light on aspects of the ra- tional-choice enterprise, which involves emotions, social norms, and institutions, thus emphasizing the empirical side of the field (Jon Elster, "Some Unresolved Problems in the Theory of Rational Behavior," *Acta Sociologica* 36 (1993): 179–90.

meanings: land, as the embodiment of natural resources in the broadest sense, captures the significance of the bounty of the earth in a particular geographical location. The existence of fertile soil, minerals, forests, or water would be considered by urban economists and geographers as significant inputs to, and factors of, production, which would have bearing on the economic development of a city and the shaping of its land market. Additionally, urban economists and geographers consider land more specifically as one of the four factors in the supply side of urban growth along with labor, capital, and entrepreneurship.[2]

Urban economics and the theory of land and land rent provide a normative framework for explaining how locational decisions are made, how the market comes to rationalize the varying land prices, and how this pricing of land assists in the segregation of land uses. Classical locational theory, as interpreted by William Alonso, suggests that each urban site expresses two commodities: location and land. Urban activities seek to maximize their utility by gaining access to a central location. All cities have central business districts providing retail and service functions to the city. Since the geometry of the city does not allow for the stacking of all activities in the sought-after CBD, a competitive price system emerges to order the urban activities in and around the CBD according to their ability to meet the market price for each location and parcel size.[3]

In a perfectly competitive environment, it is assumed that people hold productive capital and land assets because of the returns that these yield or can potentially yield in the future. Owners, therefore, seek to maximize returns by assigning to the land its highest possible use. The value of land can be increased by improvements, such as the input of capital and labor in improving drainage, installing electrical, sewage, and telephone infrastructure, and ultimately erecting buildings. The use of properties is, in theory, determined by their distance to the central business district, assuming the need of the user to have access to this CBD.[4]

Consideration of supply and demand is essential in order for planners and owners to discern the best use for each plot of land. Market prices are used as rewards to encourage input owners to find the best

2. Wilbur R. Thompson, *A Preface to Urban Economics* (Baltimore: Johns Hopkins Press for Resources for the Future, 1968), pp. 56–59.

3. William Alonso, "A Theory of the Urban Land Market," *Proceedings of the Regional Science Association* 6 (1960): 149–58.

4. General concepts appear in Edwin S. Mills and Bruce W. Hamilton, *Urban Economics*, 3d ed. (Glenview, IL: Scott, Foresman and Co., 1984), pp. 67–128.

uses.[5] In a perfectly competitive market, and in a single-CBD urban setting assuring equitable access to all, land uses and their initiators will be ordered in concentric circles around the CBD, with those willing to pay progressively higher land rents located closer to the CBD. Improvements (capital investments such as the construction or renovation of buildings) will be made to plots of land with a view to maximizing the marginal product of the input.[6] Land uses will aggregate in and around the CBD, creating specialized zones and sectors exhibiting permutations of compatible and complementary business activities.[7]

This model has a very important implication for understanding land consumption and use. One assumption, which we can readily accept for Brussels, is that access to the CBD is valuable. Traditionally, access to land in the CBD is considered important because of the possibility of agglomeration economies (economies of scale) and because of the CBD's export facility, both especially significant for manufacturing firms. A second assumption of the model is that land buyers/users have complete access to information about opportunity and make their location choices on the basis of availability and price of land and inputs such as capital and labor. A third, implicit assumption is that the inert urban fabric made up of vacated or inefficiently used buildings in the CBD will succumb to market pressure and allow new building forms and land uses to replace ones that do not produce the highest possible land rents.

A fourth and last assumption of the model involves the recognition of the market as a dynamic, integrative mechanism for the betterment of society. In spite of the level of abstraction that it can achieve and the rhetorical distance from political ideology that it maintains, neoclassical economic theory would recommend that free markets are best for society, and the more free the markets the better off society is. Markets are shown as socially and ethically noncontentious arenas where, according to Milton Friedman, one witnesses voluntary coop-

---

5. Ibid., p. 77.

6. "The marginal product . . . of an input is the change in output that results from a small change in the input quantity employed." Ibid., p. 71.

7. In this brief explanation of the model, we assume firm location to be the most significant factor. According to the model, residential location follows a similar land rent rationale responding not only to the location of the place of employment (in this case the CBD) but also to the location and attractiveness of a variety of urban necessities and amenities (schools, religious establishments, daycare, recreational areas, and aesthetic commodities, such as panoramic views).

eration among citizen-consumers.[8] The state and the "dead hand"[9] of state planning should be kept to the role of enforcer of contracts and provider of security of markets. While the influence of political structures on markets is well documented and, as noted above, criticized by some social scientists, the influence of markets on political structures and the cities they govern has not been understood as well. Apart from Marxist social scientists who have been dealing with this particular flow of causality for a long time, other social scientists have neglected to take more than a superficial stock of the manner in which political behavior is modified by microeconomic decisions. Peter Rimmer gives a powerful account of the influence of Japanese construction companies on Australia's urban structure. Eager for foreign direct investment and its collateral political benefits, local officials in Australian cities have been willing to qualify planning and zoning regulations to satisfy Japanese investors. By reordering the urban management priorities they would fail to serve at least some of the socioeconomic and environmental imperatives of their general urban development agenda.[10] Within any CBD, the economic stakes will be higher than anywhere else in the city, and within a world-class city, the pressures and potential benefits of yielding to international investors may far outweigh the resentment of small constituencies of local residents. Markets then *are* contentious mechanisms, and neoclassical economic theory by itself does not provide an explication or a remedy for the political, social, and ethical impacts they bring to bear on urban populations.

Since the supply of land is inelastic in the CBD, a number of factors typical of both American and European CBDs influence rents and the congestion of firms: the relative growth of the CBD and the composition of the market will be affected by shifts in the international economy and specifically the changing industrial mix that articulates the city economy. Changing technologies, such as advances in multimedia telecommunications, alter the labor requirements of these industries in terms of size and specialization. Both the demand for and the supply of specialized labor are significantly elastic. Public transportation, public investment in road construction, and municipal

---

8. Milton Friedman, *Capitalism and Freedom* (Chicago: University of Chicago Press, 1962), chap. 1.

9. Stuart Butler, *Enterprise Zones: Greenlining the Inner Cities* (New York: Universe Books, 1981).

10. Peter Rimmer, "Japanese Construction Contractors and the Australian States: Another Round of Interstate Rivalry," *International Journal of Urban and Regional Research* 12, no. 3 (1988): 404–24.

and private construction of parking facilities, or the lack of the above, enhance or hamper access to and within the CBD and influence the level of costs. Finally, the CBD is affected by competition from alternative urban and suburban locations.

Where urban models fall short is in reconciling the conceptual world of perfect competition with a real world where numerous structural controls of a political and administrative nature influence the market process. Most important among these are zoning regulations; fiscal policy; industrial, housing, and transportation subsidies; rent controls; conservation; and, at a higher plane, consideration of national and Regional interests.

Brussels and the quartier Européen-Léopold do not fit easily either in this simple urban economic model or in more sophisticated ones that include the consideration of multiple industries, multiple CBDs, decentralized employment, or "open city" considerations (interurban migration). The land market and private investment have clearly been of undisputed significance in the shaping of the nineteenth-century city and in the transformation of the quartier in question into a specialized CBD. An understanding of the land-rent rationale utilized by real-estate developers and private investors, as well as by the government of the Brussels-Capital Region, in their maneuvering in the market, is essential to our understanding of this peculiar CBD's formation.

The quartier Européen-Léopold, as a CBD, houses an information market and a client culture. It is the communication center for industries servicing the European Communities. The size and character of the market shapes the size and character, including the physical requirements, of the CBD. The institutions of the EC are producers of economic and social policies (and quite soon foreign policy), and a variety of service-oriented firms, such as lobbying and law firms, act as consumers and processors of the information. More important, these are bureaucratic clients who seek to influence the policies of the European Commission and Council of Ministers. Face-to-face communication is of the essence. Nothing less than prestigious, although not necessarily large, front-office operations will suffice. As a result of European Community advances toward economic integration, the quartier's centrality is rapidly increasing, and the quartier has become a coveted location for corporate investors.

This is not, however, a perfectly competitive market. The land purchasing and office building activities of mainly large British development firms in the nineteenth-century quartier Léopold in the late 1950s, just east-southeast of the then recently completed Belgian

Cité Administrative, were a straightforward response to the need for a modern office sector in a city with limited business office infrastructure. With minimal zoning regulation, developers with good political connections would purchase entire street blocks in the quartier Léopold. Prized neoclassical mansions were quickly leveled and replaced by modern multistory office blocks (see the series of air photographs of the quartier from the Institut Géographique National, 1950, 1956, 1969, 1979, 1988, and 1991). Any single building or group of buildings, regardless of historical significance or architectural value, was a potential target.[11] Density rose as this node of the Brussels CBD began to develop because rising land values induced developers to substitute away from land in the production of office buildings.[12] Until the royal decree of 28 November 1979 set forth the broad urban planning guidelines of the Brussels Agglomeration Sectoral Plan, the quartier was being dismantled piecemeal by the private sector.

The 1979 Sectoral Plan, born of the efforts of conservation-minded interest groups, signifies the first serious effort on the part of the national government to regulate the real-estate development market. Since then zoning regulations, public opinion, subsidization of housing, rent controls, and the protection of buildings of historical and aesthetic value have relatively constrained the market's impact on the remainder of the nineteenth-century city. The process initiated in the 1950s of cataclysmic consumption of the nineteenth-century neoclassical fabric of the quartier has recently been distilled into an intricate bargaining process the battlefield of which is often a single house. This process bears little resemblance to the American CBD planning tradition, where the maximization of returns is of almost exclusive importance.

A comparison between pure change in a "green fields" setting and restricted change in a mature Western European city would pro-

---

11. What was happening in the quartiers Léopold and Nord-Est was typical of the city as a whole. In spite of the outcry at the International Congress of Architects in Venice and a plea to the Belgian government to save the building, public indifference to conservation caused the 1964 demolition of an art nouveau treasure, the Maison du Peuple of Victor Horta, to build a high-rise in its place. Franco Borsi and Paolo Portoghesi, *Victor Horta* (Brussels: Vokaer, 1977), p. 62.

12. Mills and Hamilton, pp. 110–12. The elasticity of substitution of land for capital and labor, and therefore the building of mid- and high-rises in the place of townhouses, is, of course, a function of the relative cost of capital and labor. As land values rose at the quartier Léopold, developers took advantage of the lack of zoning restrictions, low interest rates and the presence of a newly arrived immigrant population of south Europeans and North Africans.

vide a very powerful image describing the structural constraints in place in Western European cities. If the quartier Européen-Léopold were a new "Ville Européenne" lacking a nineteenth-century "Léopold" component as well as its Brussels environs, the land-use model would be quite effective as a predictor of the morphology and social structure of this superspecialized city with the Continental CBD. We can imagine that it would resemble, say, corporate Stamford, Connecticut, more than any Belgian city. In a certain respect, it is surprising that the European Community administrative park was not founded as a green fields development outside Brussels. The European Community has clearly expanded far beyond what was imagined by the planners whose decision it was to locate the Community in Brussels.

In conclusion, land-rent considerations are important to the understanding of the quartier. They are the basis of calculations by the large corporate players in the quartier and the Regional government. It is clear, however, that the structural constraints are many and that land-use decisions are often influenced by political expediency, bargaining, and the intrinsic value of the historical urban landscape.

Urban Morphology

The idea that urban form encapsulates the character and qualities of urban communities in "a detailed interplay of land subdivision, building types and their combined patterns of use" is central to the field of urban morphology.[13] As the name suggests, urban morphology is the study of urban form, but it goes beyond measuring form and pattern and explores the processes and agencies that bring about the transformation of urban form, and finally the relationship between culture and urban form within a historical context.[14] The challenge to be met by urban morphology lies in understanding the links between cultural referents, historical developments, ideological

---

13. Michael P. Conzen, "Town-Plan Analysis in an American Setting: Cadastral Processes in Boston and Omaha, 1630–1930," in T. R. Slater, ed., *The Built Form of Western Cities*, p. 144.

14. "In urban design, the term is principally used for ' . . . a method of analysis which is basic to find[ing] out principles or rules of urban design' (Gebauer and Samuels, 1981); although they also note that the term can be understood as the study of the physical and spatial characteristics of the whole urban structure: this is closer to the geographer's usage." Peter J. Larkham and Andrew N. Jones, *A Glossary of Urban Form* (Leicester, UK: Historical Geography Research Series, no. 26, 1991), p. 55.

movements, and political structures and their expressions in space and particularly in the built environment. Urban morphology is therefore necessarily historicist, culturally informed, and complementary to social-scientific and humanistic views of the city.

Europe, and specifically the German-speaking world, is the original home of urban morphology. Research on the growth and structure of European continental and British cities has created a basis for comparisons, and suggests that European cities have certain common characteristics which set them apart from cities developed elsewhere. This common European cultural heritage is reflected in the literature, which strongly suggests the long-standing diffusion of urban planning and iconographic ideas throughout Western Europe.

In German geography, the townscape *(Stadtlandschaft)* and its development through history were the subject of an urban geography that was often generally expressed within the context of cultural history *(kulturhistorische Stadtforschung),* and that exhibited great devotion to historical processes and to the reconstruction of historical landscapes.[15] Its trademark emphasis on the careful mapping of the building fabric, building age, and land use as a background to plan interpretation gave rise to numerous monographs on German cities during the 1920s and 1930s.[16] Specifically, the German tradition emphasized

> the intensive and accurate observation of geographic phenomena both in the field and in maps, and searched for the processes producing such phenomena and the underlying forces involved. It sought the unambiguous conceptualization of observed phenomena on the basis of these processes and forces and expressed a readiness for testing and improvement by comparative study. It devised an appropriate cartographic expression for concepts formed and, finally, maintained an interdisciplinary approach on any geographical problem.[17]

Flourishing in the 1920s and 1930s in central Europe, it was sev-

---

15. Ibid, pp. 24–25.

16. T. R. Slater, "Urban Morphology in 1990: Developments in International Cooperation," in T. R. Slater, ed., *The Built Form of Western Cities,* p. 5.

17. J. W. R. Whitehand, "Background to the Urban Morphogenetic Tradition," in J. W. R. Whitehand, ed., *The Urban Landscape: Historical Development and Management: Papers by M. R. G. Conzen* (London: Academic Press, Institute of British Geographers, Special Publication no. 13, 1981), p. 9.

eral years before this tradition was introduced to the English-speaking geography community, as was the case with Walter Christaller's central-place theory.

Since the transplantation of the field of urban morphology in Britain, and its elaboration by M. R. G. Conzen, a triad of morphological elements (ground plan, building types, and land use) has become the query basis for such studies. In a seminal 1960 study of the English market town of Alnwick, Northumberland, M. R. G. Conzen[18] established the basic framework of these three elements in urban morphology, along with a full line of enduring morphological concepts, such as fringe-belt development, the morphological period, and most notably, the conceptualization of the burgage cycle of development.[19] His study adopted for the first time in Anglo-American geography a thoroughgoing evolutionary approach, recognized the significance of the individual lot as the fundamental unit of analysis, used detailed cartographic analysis in conjunction with field survey and documentary evidence, and conceptualized developments in the townscape.[20]

Since that time, urban research by geographers such as Peter Hall and Brian Berry, and economists such as William Alonso, focused primarily on the functional nature of cities and systems of cities.[21] In the United States, urban morphology remained largely unexplored

---

18. M. R. G. Conzen, "Alnwick, Northumberland: A Study in Town Plan Analysis." *Institute of British Geographers*, publication no. 27 (1960).

19. Burgage or burgage plot is defined as "[t]he urban strip-plot held by a burgess in a medieval borough and charged with a fixed annual rent as a contribution to the borough farm." Ibid., p. 123.

20. J. W. R. Whitehand, "Background to the Urban Morphogenetic Tradition," p. 12.

21. William Alonso's important work *Location and Land Use: Towards a Theory of Land Rent* (Cambridge, MA: Harvard University Press, 1964), was built on the Löschian tradition of the economics of location and essentially parented the urban economic tradition. Brian J. L. Berry and John D. Kasarda, *Contemporary Urban Ecology* (New York: Macmillan, 1977), present in detail the theoretical and methodological linkages between human ecology and urban geography. Brian Berry's interest in the refinement of central-place theory, the geography of economic systems, interurban networks, and Regional development has directed him to examine, in his own words, "cities as systems within systems of cities." B. J. L. Berry, "Cities as Systems within Systems of Cities," *Papers and Proceedings of the Regional Science Association* 13 (1967): 147–63. Manuel Castells's *The Urban Question: A Marxist Approach* (Cambridge, MA: MIT Press, 1977); and David Harvey's *The Urbanization of Capital: Studies in the History and Theory of Capitalist Urbanization* (Baltimore: Johns Hopkins University Press, 1985), are representative works of the critical Marxist school focusing on uneven development, urban social problems, and the capitalist dependency of cities. Some aspects of these different approaches are elaborated in this chapter.

territory, straddling the conceptual boundaries between geography, history, urban design, and art.

The field has since the 1960s been pushing beyond its traditional perimeter. An interest in relating the nature of the built environment to a cultural system of symbols[22] and social parameters has driven the field away from the contemplation of the evolution of physical aspects of the urban environment alone to investigations of the meaning conveyed by buildings, their contents, and their inhabitants,[23] and the relationship between urban form and economic function.[24] Three themes have hence been developed in research that has attempted to integrate social and humanistic aspects with a knowledge of the built fabric of cities:

- The sociotopographical reconstruction of cities
- Contextual studies exploring the wider framework of society and national economy and their effects on the built form of the city
- Investigation of the agents of change in the townscape, with an emphasis on the processes of plan formation and transformation[25]

This agenda is being explored through a new set of concepts of city expansion and evolution, most important among them being the concept of fringe belts. Described by M. R. G. Conzen as "belt-like zone[s] originating from the temporary stationary or very slowly advancing fringe of a town and composed of a characteristic mixture of

---

22. An example is Paul Wheatley's cross-cultural look at the city as an organizing principle of culture and the city as symbol. See his *City as Symbol: An Inaugural Lecture Delivered at University College, London, 20 November 1967* (London: H. K. Lewis & Co. for the University College, 1969); *Nāgara and Commandery: Origins of the Southeast Asian Urban Traditions* (Chicago: University of Chicago Geography Research Paper nos. 207–8, 1983).

23. An example would be Amos Rapoport's *The Meaning of the Built Environment: A Nonverbal Communication Approach* (Tucson: University of Arizona Press, 1990). Rapoport approaches the meaning of the built environment from the perspective of studies concerned with the developing of an explanatory theory of environment-behavior relations.

24. Examples include Anthony D. King, *Global Cities: Post-Imperialism and the Internationalization of London* (London: Routledge, 1990); and Mike Freeman, "Commercial Building Development: The Agents of Change," in T. R. Slater, ed., *The Built Form of Western Cities*, pp. 253–76.

25. T. R. Slater. "Starting Again: Recollections of an Urban Morphologist," in T. R. Slater, ed., *The Built Form of Western Cities*, pp. 10–12.

land-use units initially seeking peripheral location,"[26] fringe belts accent in the landscape the economic, political, and planning conditions that bring construction and reconstruction to urban places. Described by Whitehand as initially coarse in texture, since they usually accommodate urban uses requiring significant amounts of space (cemeteries, schools, hospitals, military compounds), fringe belts mature once further building developments have initiated the process of closure. The land uses of a fringe belt carry the seeds of its modification: the fringe belt does not passively reflect the social and cultural setting which gave rise to it, but actively influences urban developments around it, as well as its own future shape and character.[27]

This pulse-like cyclical process of fringe-belt development and modification is important in understanding the European city and our Brussels quartier. In addition to commonly observed fixation lines, or ancient fringe belts, resulting from long-standing fortifications, nineteenth- and twentieth-century fringe belts prescribe the initial development and recent expansion of the European Community administrative park. A note of caution is however necessary. Although the European Community institutions may appear in form to occupy a fringe belt, in reality neither the conditions of their placement in the quartier nor their current relationship to the city allows them to be explained in terms of traditional fringe-belt theory. As an institutional client in the city, the European Commission initially leased from the Belgian government a relatively modest amount of land in what was essentially a mature, planned residential community, and outbid the Convent of Berlaymont (a typical fringe-belt land user), which then had to move outside the city limits. Impacted primarily by world political developments among Community members, and only secondarily by business and building cycles, the site began to assume the *shape* of a fringe-belt, hosting specialized administrative and business uses slicing through the residential fabric of the quartier, without being in the fringe of the urbanized area. It is important to understand how one spore of an administrative headquarters building in the site of the Convent of Berlaymont has given birth first to an institutional "fringe belt," which in turn has become a new node of the city and a land predator. It is necessary to consider whether the EC administrative park

---

26. M. R. G. Conzen, p. 125.

27. J. W. R. Whitehand. *The Changing Face of Cities: A Study of Development Cycles and Urban Form* (Oxford: Basil Blackwell, 1987), pp. 76–94.

indeed constitutes a fringe belt of sorts or is something altogether different.

The historical dimension is of pivotal importance to understanding fringe-belt development and urban morphological studies which deal with the evolution of the morphological frame (the ground plan, building types, and land-use patterns). Unlike urban economic theory and sociological studies, urban morphology has a fundamental stake in investigating the urban fabric, both in its enduring, contemporary form, and in its previous incarnations now swept away by technological change, new functional requirements, demographic transformation, and the evolving organization of urban society. Thus, even in a contemporary topic such as the investigation of the quartier Européen-Léopold, the focus is not on present form alone, or even on the relics of previous morphological periods. The urban fabric that has disappeared is often, as in this case, as significant as the forms that remain.

The urban morphological tradition is important to this study, to the extent that it provides material evidence for the urban strategies and games carried out by quartier agents. The exploration of the material record begins by tracing the evolution of Brussels through three morphological periods, or cultural periods, each of which exerts a distinctive morphological influence upon the city and the quartier in particular.[28] During each of these periods, sets of processes relating to the distribution of cultural groups in the Region, the state of economic development and political organization of Europe, and the particular circumstances of urban management in the city, gave rise to urban forms that linger in the 1990s cityscape and play an important role in the present and future character of the city and the quartier. These morphological periods are identified as Flemish/dynastic, "French"/nationalist, and international/supranational and are elaborated in the following chapter.

Of these three periods, the "French"/nationalist and contemporary international/supranational periods are the most significant. Regrettably little has survived of the Flemish/dynastic urban environment, and its expunging during the nineteenth century is reflective of new processes at work, which I shall address in chapter 3. By contrast,

---

28. Defined as such by J. W. R. Whitehand in *Rebuilding Town Centres: Developers, Architects, and Styles* (Birmingham, UK: Department of Geography, University of Birmingham Occasional Publication no. 19, 1984). Also, defined in Larkham and Jones, *A Glossary of Urban Form*, p. 55.

the "French"/nationalist period is important to a contemporary study of the quartier because until approximately 1960 the quartier was very much a nineteenth-century and early twentieth-century artifact. On this elaborate nineteenth-century canvas, a new set of morphological processes have been at work and have given rise to a new type of CBD.

Morphological studies dealing with the evolutionary cycle of urban areas, the significance of authorship of building projects, and the locational choices of the service sector are important to this study. They instruct on the importance of consistent cartographic coverage, and on the utility of building permits in providing great detail about form, tenure, and value. Of these, the study of the evolutionary cycle of urban areas falls into the core subject of urban morphology.

An example of a study of a changing urban landscape in the M. R. G. Conzen tradition is Marek Koter's research on the morphological evolution of the nineteenth-century downtown of Lódz, Poland.[29] Through the careful examination of cartographic sources, Koter investigated the evolving morphology of Piotrkowska Street, the main thoroughfare of Lódz from 1827 to 1973. He traced the transformation of the street as a three-phase development cycle: an *institutive* phase (1827–53), during which regular urban blocks were founded on both sides of the street; a *repletive* phase (ca. 1860–1937), during which the original regular urban blocks were slowly developed and filled; and finally a *recessive* phase (1960s–73), which saw the demolition of old buildings, and was followed by a fallow period. A new cycle, no longer based on the medieval burgage plot, succeeds the fallow period.

While the quartier's urbanism is quite different, similar developmental phases are apparent in each morphological period. During an institutive phase (ca. 1840–75), the ground plan of the quartier was laid out and land was parceled out and sold. Between 1875 and 1930, land allocation on a grand scale was completed, and the repletive phase began the erection of buildings in most of the lots.[30] The Great

---

29. Marek Koter, "The Morphological Evolution of a Nineteenth-Century City Centre: Lódz, Poland (1825–1973)," in T. R. Slater, ed., *The Built Form of Western Cities*, pp. 109–41.

30. As I noted in chapter 1, the quartier Européen-Léopold consists of the greater part of the nineteenth-century quartier Léopold and quartier Nord-Est. The quartier Léopold was founded in the 1840s, while the quartier Nord-Est, more distant from the ancient Pentagon city, was developed in the 1870s. There is clearly a lag of about twenty-five years in the succession of the institutive and repletive phase in the case of the quartier Nord-Est.

Depression, the German occupation of Belgium, and the subsequent reconstruction brought an early fallow period to the quartier. The establishment of the quartier as the site for the development of the European institutions brought about the recessive phase, which was not followed by a fallow period but by a new developmental cycle based on the rapid demolition of the nineteenth-century buildings, the consolidation of the original lots, and the erection of modern structures.

In this investigation, the land economic perspective contributes to our understanding of the quartier as an area in morphological flux and perhaps as a variant fringe belt. Land-rent theory provides an understanding of business and building cycles that drive the physical growth of the institution and that combine with other land users to determine the bidding process.

Where the morphological tradition appears to fall short is in the consideration of agency in more than outline form. To illustrate the significance of "interactions between main parties," Whitehand accounts microscale bargaining cases between developers and planning authorities in Devonshire, Hyrons, Downside, and Parade in the United Kingdom, presenting the interactions between the agents as "the framework within which an outcome in the landscape is to be understood."[31] In this context, developers and planning officers function as antagonists in a contest where the stake is integrated, conservation-sensitive urban landscape development. Although the particular elements of the interaction are described well and in detail, Whitehand does not elaborate on the nature and principals of the decision-making process—other than that it was largely based on economic considerations. He should be credited, however, for pointing out the need for further study of this type of interaction. That is where notions of rational choice—often drawing on the optimizing and utility-maximizing nature of human interrelations within a free market system—and regimes may complement Whitehand's initial explorations of decision making in urban landscape management.

These sketches of land-rent theory and urban morphology are indicative of the interdisciplinary requirements of the study. The listing is not exhaustive but rather introduces two of the theoretical traditions represented in the study. As already noted, rational-choice and regime-related theoretical considerations will be best treated in

---

31. J. W. R. Whitehand, *The Making of the Urban Landscape* (Oxford: Blackwell Publishers, 1993), p. 177.

the substantive chapter on political process. Works on the structure of world-class cities, on market centers and the dynamics of office location, as well as on architectural and planning theory also support the conceptual framework of the study.

3

The Historical Background

O N 1 MARCH 1502 Philippe le Beau, father of Charles V, elevated Francisque de Tour et de Tassis to the captainship of the French Kingdom's postal service. De Tassis would establish himself in a beautiful manor close to the Church of Notre-Dame-au-Sablon and would organize the central node of a postal network in Brussels. With the accession of Charles V to the Habsburg throne, Brussels became the imperial postal entrepôt. A working register of the period lists the following service routes and estimated delivery times, and attests to the centrality of the city in dynastic politics and economics:

Brussels-Paris: forty-four hours during the summer; forty-eight hours during the winter

Brussels-Blois: two days and a half during the summer; three days during the winter

Brussels-Lyon: four days during the summer; five days during the winter

Brussels-Granada: fifteen days during the summer; eighteen days during the winter

Brussels-Toledo: twelve days during the summer; fourteen days during the winter

Brussels-Innsbruck: five days during the summer; six days and a half during the winter[1]

Almost six hundred years later, economic prosperity and changes in transportation technology have caused the horses and carriages of Valois France to be replaced by the automobile and soon the TGV su-

---

1. Original sixteenth- century royal edict quoted in Jo Gérard, "L'insolite: Bruxelles, capital historique des postes," in *Télé-Bruxelles raconte Bruxelles* (Braine l'Alleud: J.-M. Collet, 1991), pp. 263–66.

perfast trains. The same postal itineraries today are reached from Brussels in as many hours as days in the time of de Tassis's tenure.

Before the centrality of Brussels in Western Europe was acknowledged in this appointment, however, Brussels historically derived its raison d'être from its felicitous location straddling important east-west trading routes and a number of waterways between Paris, Amsterdam, Cologne, and the Flemish ports on the North Sea. Thus, the initial justification for the tenth-century settling of the city may very well have been economic. Like the more ancient cities of London and Paris, Brussels was situated along a river, at the furthest inland point where the river was still navigable. Although before 1830 the city of Brussels was territorially limited to the so-called Pentagon city, and although the quartier Léopold is a distinct nineteenth-century extension of Brussels, the quartier's historical antecedents in the Pentagon must be understood in order to understand the historical and material frame that today is being negotiated out of existence for the sake of the European Communities and EC-related businesses. An appreciation of this ancient urban fabric will allow us to make an educated assessment of whether the radical changes in quartier Européen-Léopold of the 1990s constitute a disaster or a functional necessity.

The framework used here is influenced by Elisabeth Lichtenberger's model of European historical city types. In "The Nature of European Urbanism," she points to historical continuity in the Western European urban tradition, determined by a distinct political structure. Consequently, she distinguishes four historical city types, which dominate the European landscape from the medieval to the modern era: the medieval burgher city of the feudal territorial state, the city of the nobility as a creation of the absolutist state, the industrial city of Liberalism, and the New Town of the social welfare state and socialist system.[2] Lichtenberger's useful family tree of European cities models urban change in ecological terms, counterposing adaptable and reproduceable urban centers incorporating urban and socioeconomic innovations against stagnant, "one period" cities. Brussels, which belongs to the former grouping, fits well in this model. Lichtenberger's typology has been modified here to reflect the emphasis of this study on the importance of political centrality on urban form and the world economic developments of the last quarter century.

---

2. Elisabeth Lichtenberger, "The Nature of European Urbanism," *Geoforum* 4 (1970): 47.

Brussels in the Era of Dynastic Rivalry

The important connections to be made here are between the development of spatial economic linkages and trade—which proliferated
during the Carolingian era (ninth and tenth centuries) as a result of
political cohesion, the maintaining of portions of the Roman road network, and greater security in trading routes—and the growth of
urban centers taking advantage of these increased communications.
After these roads fell into disrepair during the eleventh century with
the disintegration of the Carolingian Empire, the overland Bruges-
Cologne road, connecting the North Sea coast with the cities along the
Rhine valley, brought prosperity to the cities along its path: Eecloo,
Ghent, Alost, Brussels, Léau, Saint-Trond, Maastricht, and Aachen.[3]
    The emergence of feudalism as the dominant socioeconomic sys-
tem in medieval Western Europe, rural-to-urban migration, and the
protoindustrialization of the textile sector were the key factors driving
the growth of Brussels from hamlet to city. Fundamentally important
to the urban political landscape of the Low Countries and frontier
cities such as Brussels were the dynastic rivalry and the bartering of
lands between France, the Holy Roman Empire, Spain, and later the
Habsburg Empire.
    We can visualize the kingdoms and empires of dynastic Western
Europe as a field of tectonic plates, continuously grinding against
each other, one extending its margins in a flurry of royal marriages
and diplomacy, while the margins of another are being swallowed
up by war, dowries, and family feuds. Where the major "plates"
met, there would frequently arise changes in control of the land and
in feudal relationships. Such a frontier region was that of the Low
Countries, which today is occupied by the Kingdoms of Belgium and
the Netherlands. The twelfth-century duchy of Brabant, which
largely coincides with the modern Belgian provinces of Brabant, Antwerp, and the Dutch province of Noord-Brabant, was in fact part of
the Germanic Empire, although the counts of Flanders were vassals
of the king of France.[4]
    There is little information on the founding of Brussels. Adequate
archaeological evidence exists, however, to ascertain that the area has
been under continuous occupation since the Roman period. Material

---

3. Marcel Vanhamme, *Bruxelles: De bourg rural à cité mondiale* (Brussels: Mercurius,
1978), p. 24.
    4. Paul de Ridder, *Bruxelles: Histoire d'une ville brabançonne*, trans. Emile Kesteman
(Gand: Stichting Mens en Kultuur, 1989), p. 89.

evidence suggests that the original focus of settlement was either a primitive Merovingian chapel dedicated to St. Michel (later the cathedral of Sts. Michel et Gudule) on the eastern bank of the river Senne/Zenne,[5] or a fortified military encampment on the island of Saint-Géry, one of the three naturally protected islands on the river Senne/Zenne.[6]

In 980, Brussels was introduced into the political and military balance of power of the frontier region by the Duke of Lower-Lorraine, Charles of France, with the construction of a chapel to St. Géry,[7] and a *castrum*, a riverside fortified camp, which was probably little more than a rectangular camp surrounded by earthen walls plus a single tower, a palisade, and a moat.[8] Located on three river islands—like Paris on the Ile de la Cité and the Ile Saint Louis—and surrounded by marshes, it dominated both the land and river routes. The camp was soon surrounded by houses, chapels, port facilities, and marketplaces, in accordance with the strategic doctrine of the age.[9] By the twelfth century, the military camp was moved to the eastern escarpment of the valley overlooking the river and the growing city and included a manorial residence (the ancient palace of Coudenberg, now replaced by the Royal Palace).

Each of the groups and movements active in the affairs of the city—the Catholic Church, the burgher artisans and merchants, the Regional traders, and the ruling aristocracy—gave the city physical and social aspects of its complex personality as monastic center, commercial hub, and city of leisure, as well as city of guilds, dynastic outpost, and city under successive occupations.

Eventually the small settlement extended beyond the river islands to the surrounding higher ground and was enclosed by a 4-km-long oblong circuit of walls (1063–1100), which featured fifty towers and seven gates[10]—a far cry from the simplistic earthen fortifications of the first *castrum*. These first fortifications were completed during the

---

5. P. Léfèvre, "Le problème de la paroisse primitive de Bruxelles," *Annales de la société royale d'archeologie de Bruxelles* 38 (1934): 106–16.

6. G. Despry, "La génèse d'une ville," in J. Stengers, ed., *Bruxelles: Croissance d'une capitale* (Antwerp, 1979), pp. 28–37; and M. Martens, "Les survivances domaniales du 'castrum' carolignien de Bruxelles à la fin de Moyen Age," *Le Moyen Age* 69 (1963): 641–55.

7. Martens, 641–55.

8. Vanhamme, p. 20.

9. Martens, 641-55.

10. P. Bonefant, "Les premiers remparts de Bruxelles," *Annales de la société royale d'archeologie de Bruxelles* 40 (1936): 7–46.

twelfth century. The area enclosed defined the first foci for urban growth and prescribed a space with special qualities: security from invasion (fortifications), security from floods (settlement on high ground), and subordination to political authority (again, fortifications and the Palace of Coudenberg) and to religious authority (the ancient chapel of Sts. Gudule and Michael, by that time an important church). Moreover, the construction of a sanctuary to St. Nicolas, the patron of merchants, at some distance from the river port suggests that the merchant community prospered from the feudal partnership between the overlord and the burghers and expanded its control over the lowland western part of the walled city.[11]

This separation between vernacular and official landscapes was the origin of the separation of Brussels into a "high city" and a "low city." The former became the site of most public edifices and residences of the nobility, while the latter was home to artisans, laborers, and merchants and eventually to the institutions and spaces that represented them: the first city hall (fourteenth century), the guild houses, and the specialized marketplaces.

By the end of the twelfth century, Brussels was a city of modest significance situated directly on the frontier between the kingdom of France and the Holy Roman Empire. It was known as an emerging ecclesiastical center with significant links to Rome, Valladolid, Madrid, and Vienna. Spin-off parochial hospitals, convents, monasteries, and retreats were a significant feature of the morphological and economic landscape until their suppression and exclusion from the city beginning in 1773 with the censure of the Jesuit order by the Empress Marie-Thérèse.[12]

While other Brabantine cities such as Leuven and Mechelen enjoyed the status of ducal or princely capitals, Brussels slowly attracted the manufacture of textiles, tapestries, and other luxury goods. By the thirteenth century, and most certainly during the fourteenth century, Brussels enjoyed a profound level of prosperity thanks to the success of its woven wool manufacturing. Next to the textile industry, other export industries, such as leather and metalworking, also prospered.

With population growth, the lowlands of the city were bustling

---

11. Arlette Smolar-Meynart, "L'evolution du paysage urbain," in Arlette Smolar-Meynart and Jean Stengers, eds., *La région de Bruxelles: Des villages d'autrefois à la ville d'aujourd'hui* (Brussels: Credit Communal, Collection Histoire Serie no. 16, 1989), p. 49.

12. Luc Janssens and Lisette Danckaert, "La grande propriété immobilière et son evolution," in Arlette Smolar-Meynart and Jean Stengers, eds., *La région de Bruxelles: Des villages d'autrefois à la ville d'aujourd'hui* (Brussels: Credit Communal, Collection Histoire Serie no. 16, 1989), pp. 196–211.

with activity. Building within the confines of the first line of fortifications was completely unplanned. Access between the low city and the high city was limited to three streets. Most houses were built of wood, clay, and straw. The city residences of the aristocracy and the wealthy merchants, however, were often built of stone and bore names like "Cantersteen," "Ketelsteen," and Meynaersteen" (*steen* in Dutch translates as "stone").[13]

This organic urbanism did not appear, however, to hamper the expansion of economic activity. The Grand Place was established near the port and the first *castrum* during the tenth or eleventh century and was surrounded by specialized markets: to the southwest the market for charcoal; to the southeast the markets for poultry and cheese; to the north, the markets for tripe, used goods, shoes, butter, fish, vegetables, and leather. Further to the southeast and northeast, and close to the fortifications, were located the markets for horses and livestock, which had greater space requirements. The fragrances associated with such mercantile activity were far from the residences of the nobility on the eastern escarpment overlooking the city.

By the thirteenth century, closed markets (*halles*) were being constructed in the midst of the open air markets. By the fourteenth century these became more numerous, and in the fifteenth century, the city built a significant and architecturally elegant closed market for its famous textile products. Thirteen water mills along the banks of the Senne river provided flour millers, beer makers, tanners, and metalworkers with mechanical energy.[14]

The fourteenth century ushered in the Golden Age of Brussels. For the next two centuries, the city was a premier center for art and architecture in the Low Countries. The fourteenth century saw the building of numerous churches and the nascence of a long tradition in the construction of public buildings with administrative or political functions. Claus Sluter moved to Brussels from Haarlem, Holland, to register with the guild of stonemasons and to bring together the group of artisans and architects who would replace the dominant French gothic style in sculpture and architecture with a new gothic style with a distinct Brabantine flavor.[15] This artistic renaissance reflected more than economic prosperity. In the political arena, it signified the fruition of a century's worth of civic warfare for burgher rights and greater autonomy from the ruling nobility. The struggles of a primi-

13. de Ridder, p. 18.
14. Smolar-Meynart, pp. 62–63.
15. de Ridder, pp. 18–19.

tive urban proletariat drawn from the class of weavers helped realize citizens' rights codified in the Charter of Cortenberg (1312), the Chartes Romanes (1314), and the Joyeuse Entrée, which remained in force until the 1790s.[16]

By the last quarter of the fourteenth century, the growth of the population necessitated the erecting of a new, pentagonal 8-km-long set of fortifications enveloping the twelfth-century *castrum*. It was during this century that the city's magistrates—now with greater authority over the affairs of the city—began a campaign of embellishing the city. Serious efforts were made to regularize the ground plan of the riverside marketplace. A special administration for public roads was created and two Maîtres de la Chaussée were assigned the task of paving the streets and creating a plan for the downtown, to allow for the enlargement of the marketplace and the designing of a rectilinear central square. By the end of the fourteenth century, space was created between the rue de la Colline and the rue des Harengs for what is known today as the Grand Place. The many houses occupying that space were expropriated en masse and demolished.[17]

By 1430, the city fell under the jurisdiction of the House of Burgundy, which was related to the House of Valois of France. The city became one of the seats of the traveling courts of the Burgundian nobility, although not the capital of the kingdom. Burgundian rule coincided with the decline of the city's traditional wood-trading and textile industries caused by English competition. This decline resulted in a long period of economic depression that left large areas of the city in great disrepair, and the population unemployed and famished. The accounts of people dying of hunger and cold in the streets of Brussels are confirmed by the records of the monastery of St. Eloi, which made extensive attempts to alleviate the hardship of the Bruxellois during the winters of 1451 and 1452.[18]

Interestingly, this period of economic hardship did not appear to restrict plans for the growth and embellishment of the city, especially around the Grand Place. Funds for building projects were provided by the receipts from the itinerant Burgundian Valois court and by the gradual emergence of the tapestry industry in place of the traditional textile industry. The political will and necessary organization were

---

16. John Fitzmaurice, *The Politics of Belgium: Crisis and Compromise in a Plural Society* (London: C. Hurts & Co., 1988), pp. 8–10.

17. G. Des Marez, *Guide illustré de Bruxelles: Monuments civils et réligieux* (Brussels: Touring Club Royal de Belgique, 1979), pp. 36, 37.

18. de Ridder, p. 27.

the result of the efforts of the Valois to consolidate and organize their holdings into a centralized system of governance, without violating the royal obligations established by earlier charters and by the customary rights of burghers, which were made even more secure after the popular uprising of 1421 against Jean IV and Jaqueline de Bavière.[19] By 1454 the City Hall, conceived in a sumptuous gothic style, was finally completed. The guilds around the Grand Place had triumphed against Jean IV and proudly planned their own opulent houses. As G. Des Marez reports, the guild of the glovemakers was so proud of the Grand Place guild house it was constructing in 1464 that in order to gain certain advantages from the city's magistrates, it was assuring them that the construction was going to be "a jewel of the city."[20]

With the ending of the Valois lineage, Brussels and the Low Countries came under the jurisdiction of the Habsburgs of Spain. Under the reign of Charles V (1515–55) Brussels became not only the princely capital of the Netherlands, but also the capital of the Habsburg Empire. It was under his rule that the family of Tour et Tassis organized the international postal service mentioned earlier in this chapter. Also at this time Brussels was visited regularly by Europe's most powerful monarchs, as well as by ecclesiastical and secular dignitaries. With increasing prestige and political centrality, however, came the curtailment of burgher autonomy and the increased persecution of non-Catholics. Brussels became one of the battlegrounds of the Reformation, claiming its share of victims of religious persecution. Under the reign of Spanish-born Philip II, the son of Charles V, the Brussels court was inundated with Spanish nobility at the expense of the local gentry. In 1557, of 1,500 courtesans, no fewer than 1,300 were Spanish.[21] The Hispanicization of the Brussels court was reinforced by several Catholic monastic constituencies that established themselves in the Pentagon: Jesuits in 1586, Capuchins in 1587, Augustines in 1589, Riches Claires in 1595, Thérésiennes in 1606, Carmes in 1612, Minimes in 1618, and Brigittines in 1623.[22]

Elements of the physical aspect of the sixteenth-century city are preserved in a number of maps and engravings. One of the most ac-

---

19. Renée Doehaerd et al., *Histoire de Flandre: Des origines à nos jours* (Brussels: La Renaissance du Livre, 1983), pp. 94–95.

20. "[U]n joyau de la ville." Des Marez, p. 37.

21. de Ridder, p. 27.

22. Pierre Mardaga, ed., *Le patrimoine monumental de la Belgique: Bruxelles*, vol. 1, book A, *Pentagone A-D* (Liège: Solédi pour les Communautés Française et Flamande, 1989), p. xlii.

curate maps of Brussels, dated 1550, belongs to a series of two hundred fifty manuscript maps produced by Jacob van Deventer (ca. 1500–75)[23] and commissioned by the Emperor Charles V and King Philip II (fig. 5). While Deventer's map is not sufficiently detailed to provide information about habitations, it clearly shows the extent and nature of Brussels's urbanism at that time. The city had already spilled over the first line of fortifications into the space enclosed by the second line of fortifications, but building remained to a certain degree contained along several nonrectilinear road axes: rue Haute north to the Cathedral of Sts. Michel et Gudule, and along its extensions to the Portes (gates) de Schaerbeek and Leuven; and in an east-west direction, rue de Flandre, Sainte-Catherine, Marché-aux-Poulet and aux-Herbes, de la Madeleine, Coudenberg, and along the road leading to the Porte de Namur. Water routes included a canal that ran along the western bank of the river Senne and was linked to the principal and secondary moats, and the port facilities in the commercial heart the city.

The most densely occupied parts of the city were the Grand Place and its environs, the neighborhoods adjacent to the river, and the Sablon and Beguinage neighborhoods. The resulting open spaces, especially to the north, west, and southwest, were available for cultivation. Houses for the thirty-five thousand inhabitants of the city were colored red, and the large buildings were distinguished by their gray roofs. The fortifications were complemented by towers with mansard roofs. While Deventer worked on a scale that did not allow the elaboration of physical characteristics of houses and public buildings, other maps of the period, one by Lodovico Guicciardini (fig. 6),[24] and another by Georg Braun and Frans Hogenberg (fig. 7),[25]

---

23. Jacob van Deventer was a medical doctor, geographer, cartographer, and geometrician educated at the University of Leuven. His two hundred fifty maps of cities of the Low Countries were produced on the basis of field research, were always oriented north, and used a consistent scale (1:9,000), which allowed comparisons of areal extent.

Jacob van Deventer, manuscript map of Brussels and its environs, ca. 1550, dimensions 81 x 67 cm, Bibliotheque Royale, Section des Manuscrits 22090, in Lisette Danckaert, *Bruxelles: Cinq siècles de cartographie* (Tielt and Knokke: Lannoo and Mappamundi, 1989), pp. 14–15.

24. Lodovico Guicciardini, woodcut map of Brussels, ca. 1567, dimensions 25 x 34 cm, Bibliotheque Royale, Section des Livres précieux VB 10056 B; engraved map of Brussels, ca. 1581–82, dimensions 23 x 31 cm, Mappamundi Collection, Knokke, in Danckaert, pp. 30–33.

25. Georg Braun and Frans Hogenberg, engraved and illuminated map of Brussels, ca. 1572, dimensions 33 x 48 cm, Mappamundi Collection, Knokke, in Danckaert, pp. 28–29.

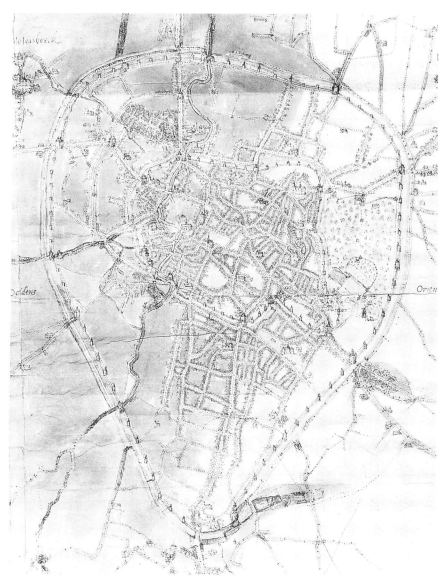

Fig. 5. Manuscript map of Brussels by Jacob van Deventer, ca. 1550. 81 x 67 cm, Bibliothèque Royale, Section des Manuscrits 22090.

reveal that the city consisted almost entirely of three- to four-story townhouses with pointed or gabled cornices, densely built on irregularly shaped city blocks, and with very little green space in the interior of these blocks. These maps make a very sharp distinction between built area and green *intra muros* space, as well as between these and the *extra muros* landscape that is dominated by neatly delineated agricultural parcels, canals. The *extra muros* landscape features a scattering of farms and the beginnings of surrounding municipalities along the roads leading out of the city gates, also distinguished in Deventer's map.

The rise of mercantilism, the expansion of the textile sector, and the improvement of textile technology brought a brief period of prosperity to Brussels, until the broadcloth industry moved westward into rural parts of western Flanders and northern France, and eastward to rural Brabant, Liège, and the lower Rhineland.[26] According to N. J. G. Pounds, the population of Brussels peaked at ninety thousand at the end of the fifteenth century, and stopped growing after political influence was transferred first to Vienna and subsequently to Valladolid.[27] The religious wars of the sixteenth century, the rigidity of the urban guilds, changes in fashion, and the growth of cloth processing in rural homesteads undermined the prosperity of Brussels and halted its demographic and urbanistic growth. Maps from the seventeenth and eighteenth centuries attest to this fact.

Almost one hundred years after Deventer's and Guicciardini's maps, the map of the city by Martin de Tailly (1640) and its approximate copy by Joan Blaeu (1649) give still greater detail about building types and the interior of the street blocks (fig. 8). De Tailly's map comprises six sheets, measuring a total of 84 x 114 cm, in addition to three sheets providing a bird's eye view of the city, and six more depicting buildings and including the legend.[28] His map and Blaeu's copy show a series of individual townhouses, churches, and public edifices in mock bird's eye view, while revealing the great extent of green space used as vegetable gardens or orchards in the interior of the street blocks. This custom of interior green space continues in the nineteenth-century extensions of the city. It is in the post–World War II period that urban densities increase to the point of absorbing the interior gardens.

During the religious strife of the sixteenth century, Brussels buildings were influenced relatively little by the pure Renaissance architectural style. The Palais Granvelle (1550–55), now destroyed, was

---

26. N. J. G. Pounds, *An Historical Geography of Europe* (Cambridge, UK: Cambridge University Press, 1990), pp. 234–35.

27. Ibid., p. 224.

28. Martin de Tailly, *Bruxella Nobilissima Brabantiae Civitas*, dimensions 84 x 114 cm, (Brussels: Bibliothèque Royale, Estampes, 1640[a], 1748[b]), in Danckaert, pp. 35–37.

Fig. 6. Woodcut map of Brussels by Lodovico Guicciardini, 1567. 25 x 34 cm, Bibliothèque Royale, Section des Livres précieux VB 10056 B.

Fig. 7. Engraved and colored map of Brussels by Georg Braun and Frans Hogenberg, 1572. 33 x 48 cm, Mappamundi Collection, Knokke.

Fig. 8. Map of Brussels by Joan Blaeu, 1649.

one of the few buildings in Brussels that reflected the classical orders and proportions adopted and interpreted by Renaissance architecture. Less paradigmatic, the main wing of the Palais d'Egmont at the Petit Sablon features first-floor rectangular windows decorated with moldings over semicircular arches. Gothic style continued to be prevalent throughout the city, with elements of the Renaissance style incorporated in the ornamentation of the façades. These features also appear in the Grand Place, where the elaboration of the classical orders reached its highest expression during the baroque period.[29]

During the seventeenth and eighteenth centuries the territory occupied by modern Belgium and the Netherlands became one of the great battlegrounds of the War of the Spanish Succession. The Treaty of Utrecht in 1713 eventually brought the Spanish Netherlands, including Brussels, to the House of Austria. Adjustments in the ground plan undertaken during the seventeenth century are still visible today: place Anneessens, situated on the traces of the rue Neuve and the Vieux Marché, retains its recognizable seven radiating streets.[30] Little remains, however, of the late medieval and Renaissance landscape of buildings after the 1695 bombardment of the city center during the War of the Spanish Succession. Danckaert reports that the legend of an Italian copy of a French map by Nicolas de Fer titled *Plan du bombardement de Bruxelles par l'armée du Roi le 13, 14, et 15 Aoust 1695*[31] details the extent of the physical damage (fig. 9).

> The heart of the city was destroyed. Some four thousand houses lay in ruins or were greatly damaged by the fire. The buildings surrounding the Grand Place were devastated . . . This ruination brought some benefit to the city: the reconstruction effort was based on strict building directives that produced the grand regeneration and embellishment of the Grand Place, allowed for the widening of roads and the general amelioration of traffic.[32]

In the aftermath of the bombardment, houses adopted a standard style which combined brick with sandstone in façades with simple lines, derived from late gothic architectural forms, and embellished with either Renaissance or baroque elements. The guild houses sur-

---

29. Mardaga, ed., p. xl.

30. Ibid., p. xxiv.

31. Nicolas de Fer, *Plan du bombardement de Bruxelles par l'armée du Roi le 13, 14, et 15 Aoust 1695* (Brussels: Bibliothèque Royale, Section des cartes et des plans XXXI, 1695).

32. Danckaert, p. 31.

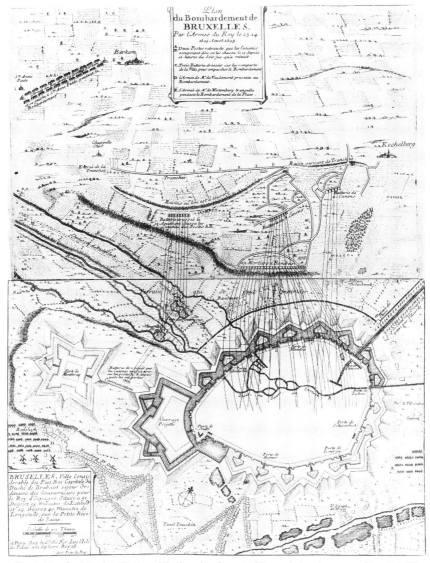

Fig. 9. Map by Nicolas De Fer of the bombardment of Brussels by the army of Louis XIV, dated 1695. Nicolas de Fer, *Plan du bombardement de Bruxelles par l'armée du Roi le 13, 14, et 15 Aoust 1695* (Brussels: Bibliothèque Royale, Section des cartes et des plans XXXI, 1695).

rounding the Grand Place were rebuilt in grand baroque style immediately after the bombardment, and some corrections were made in the rectilinearity of the Grand Place and some of the side streets,[33] but the city ground plan was not reconfigured overall. Owing to the high density of building, lots were rectangular in shape, narrow streetside, and rather long. The urban legacy of the late seventeenth-century built landscape is preserved in religious edifices, in names of the markets now borne by streets, and in small fragments of fortification.

After years of warfare and religious persecution, Brussels was to experience a period of tranquillity under the moderate rule of distant Vienna. Brussels's urbanism experienced growth under Austrian rule, although maps from this period indicate that the city had not yet completely filled the space inside the pentagonal fortifications. In a splendid map by Louis-André Dupuis (1777) commissioned by Count Joseph de Ferraris, the city still appears focused on the Grand Place, although new forms are discernible on the site of the destroyed Palace of Coudenberg, and around the park (fig. 10). The remnants of the ancient interior fortifications are still visible, while the pentagonal periphery of walls has been transformed into tree-shaded promenades.[34] Four monumental building groups were founded in the midst of the older built fabric of the Pentagon city: the place des Martyres (1774–76) designed by Claude Fisco; the place Royale (1775–81), conceived by Parisian architect Nicolas Barré to evoke the royal squares Stanislas at Nancy, and Vendôme and Vosges in Paris, and executed by Gilles Barnabé Guimard; the building groups surrounding the park (1776–85); and the place du Nouveau Marché aux Grains (1787), attributed to Nivoy and Fisco.[35] They were created in the neoclassical style, which was becoming the *style internationale* of the time, influencing the building of public and private edifices from Stockholm to Athens (fig. 11).

The neoclassical style replaced the baroque form and emphasized regularity and symmetry. French-style neoclassical quartiers arose in a city still characterized by Flemish organic planning. Their presence and the tastes of the ruling elite of the city plotted a new course for Brussels urbanism. Their influence and the influence of financial promoters and French architects continued to change the city toward a French urban vision throughout the following century.

---

33. Des Marez, pp. 39–41.

34. Louis-André Dupuis, *Plan topographique de la ville de Bruxelles et des ses environs* (Brussels: Bibliothèque Royale, Section des cartes et des plans IV, 6, 1777), in Danckaert, pp. 86–87.

35. Des Marez, pp. 111, 248–52, 292–94; Mardaga, ed., pp. xxv–xxvi.

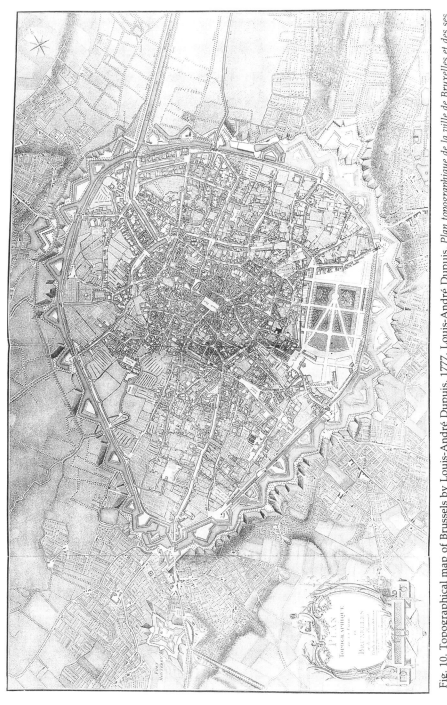

Fig. 10. Topographical map of Brussels by Louis-André Dupuis, 1777. Louis-André Dupuis, *Plan topographique de la ville de Bruxelles et des ses environs* (Brussels: Bibliothèque Royale, Section des cartes et des plans IV, 6, 1777).

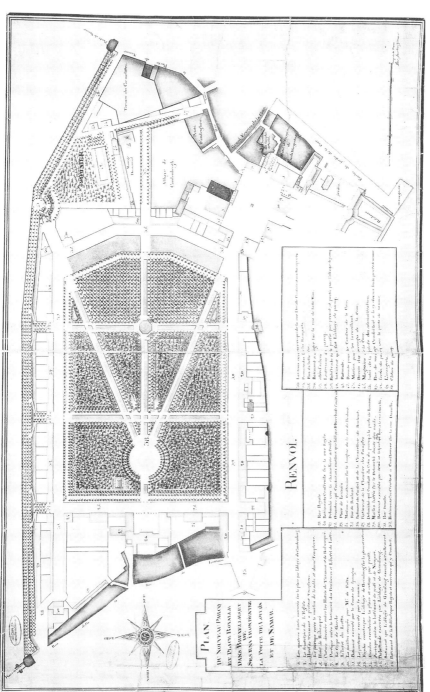

Fig. 11. Plan of the new park and place Royale by Joachim Zinner, ca. 1775.

The eighteenth-century history of Brussels was punctuated by French and Austrian dynastic military competitions for the control of the continent. With the conclusion of the War of the Austrian Succession, Brussels remained part of the Austrian Netherlands (Treaty of Aix-la-Chapelle, 1748). In 1789, the "Vonckists," a politically progressive movement seeking the introduction of a broader franchise, defeated the Austrian forces at Turnhout and captured Brussels. Their proclaimed United States of Belgium was short-lived, as the Austrians recaptured Brussels on 3 December 1790.[36] The French Revolution and the Napoleonic wars were to change the map repeatedly. After a brief period under the French revolutionaries, Austrian control was restored in 1793, only to be lost to the French again in June 1794.[37]

With the collapse of the ancien régime, Brussels gradually expanded out of its Flemish medieval and Renaissance morphological frame. When the pentagonal fortifications were demolished in 1810 and replaced by a wide tree-shaded boulevard, physical barriers to urban growth were removed, and the city was brought into competition with surrounding communities that had developed on the radial roads leading to other cities in the Brabantine hinterland.

A year after the Congress of Vienna, the Second Treaty of Paris (1815) created a greater United Netherlands, largely at the insistence of Britain, as a buffer state on the border of France. It included most of modern Belgium and Luxembourg. Fifteen years later, revolting against Dutch royal rule and the limitation of the role of the Catholic Church in public life, Belgian Liberals and Catholics united to found an independent Catholic Belgium, with Brussels as national capital. Independence brought to the city the "French" period of urbanism.

Brussels in the Era of Nationalism

In 1830, when the territories of modern Belgium became independent from the Kingdom of the Netherlands, Brussels was part of a relatively evenly balanced Regional urban network that included such centers as Antwerp, Ghent, Namur, and Liège. What these other cities lacked that Brussels had was the role of national capital. The formation of new political boundaries and the designation of Brussels as the capital of the newly independent Belgium combined

---

36. Fitzmaurice, pp. 19–21.
37. Ibid., p. 22.

to propel the city into a new sphere of urban growth.

While Brussels exhibited a distinctly Flemish urban morphological frame, civic culture, and population composition up to 1830, it was to undergo a radical refashioning during the remaining part of the nineteenth century, owing to its role as the national capital of the new Belgian monarchy. No area of the city was left untouched, while multiple frontiers of transformation were opened with a view to modernization and embellishment.

It was clear to the government of the new country that Brussels, as a national capital, would need to meet administrative needs and political imperatives by transforming itself into an indisputable icon symbolizing the nascent Belgian nationalism. It is important to note that this vision was fostered by an elite of French culture, for the most part of Wallonian origin. Moreover, new ways of thinking about urban planning (especially in mid-nineteenth-century France) had a profound effect on the civic leadership and the planners of Brussels, and this new planning philosophy became an important dimension of the national urban icon.

The need for change was undoubtedly practical; traditional, unplanned Flemish urbanism was not suitable to capital city activities. It was very difficult to accommodate the new governmental infrastructure into a city honeycombed with narrow and sinuous streets, traversed by the river and canals and lacking large open spaces inside the Pentagon. Space had to be found for a variety of ministerial and government service buildings necessary for the running of state and economic affairs, such as the diplomatic missions to the new regime, the Royal Palace, the Parliament, the courts of justice, and the stock exchange, as well as the military encampments and training grounds needed for the protection of the capital and the government.

The process of change had begun a few years before independence. Changes in the transportation infrastructure were being considered and carried out (for example, the opening of the rue Royale). Perhaps the most significant battle was fought over access inside and into the city. The urban fabric transformations cited below had a profound impact on the native Flemish population which inhabited the ancient *intra muros* city. Entire medieval neighborhoods were demolished, and most streets outside the more ancient perimeter of fortifications were swept away and replaced by planned streets. Inefficient medieval circulation patterns were gradually eliminated as wide new boulevards were cut through the ancient neighborhoods and toward new routes out of the city. Again, there were practical, economic considerations underlying the building and locating of the four Brussels train stations of Léopold, Nord, Midi, and Central. Furthermore, the building of a modern port for seagoing vessels north of the city not only became necessary with the industrialization of the city's economy, but also brought the North Sea closer to Wallonia-based industry. Finally, the most monumental infrastructure work was the 1871 subterranean diversion of the Senne/Zenne river and its subsequent vaulting with modern boulevards.

Considerations motivating these changes were political and ideological. The decision making was to no degree in the hands of the affected population. The constitution was drafted exclusively in French, and the substantial property qualifications for election suggest that the political class was to be drawn from the French-speaking bourgeoisie which had been the force behind the revolution for independence. This political group introduced to the city landscape not only a new set of government buildings and political structures but also a complete set of cultural and social institutions which were deemed important to any modern capital city: museums, an opera house, theaters, the botanical gardens, academies, and French-language schools. The effort to stamp Brussels as a French-culture capital was clear.

The variety and function of new buildings and the transformation of the city's ground plan were not the only important signs of a specific cultural imprint being introduced to the city. The architectural and artistic language employed in realizing them was itself also quite new. The architectural language quoted was drawn from established pure and eclectic models of the period which were either French or in French taste, and were therefore decidedly foreign to the preindependence cityscape. The process was not entirely new; as noted earlier, it had already started in the latter part of the eighteenth century with the neoclassical complexes by Guimard and Fisco. In practically every instance of planned extension into open space, such as the creation and annexation of the rectilinear mansion-lined quartier Léopold (1852), the opening of the Champs Elysées-like avenue Louise to the gardens of the bois de la Cambre (1864), and the creation of the French-style quartier Nord-Est by Gédéon Bordiau (1875), there was a wholesale adoption of French urban fashions in terms of ground plan and building types. Brussels was indeed only a few years behind Paris in employing the design developments characteristic of the city. Henri Moke wrote about the quartier Léopold in 1844:

(T)he quartier Léopold will eclipse with its magnificence all which Brussels has to exhibit today. There will unfold majestic avenues lined with mansions resembling palaces. In no other place is the opulence and luxury of our age displayed with greater pride. One could say that like London and Paris, Brussels wants to have its own West End or its Chaussée-d'Antin.[38]

In the interior of the city, old neighborhoods were converted into wide, radial boulevards, and squares, as in the case of the quartier Notre-Dame-aux-Neiges (1874) on the northeastern corner of the Pentagon, and were parceled off to new buyers. Alternatively, they were transformed into complexes of gardens and public buildings, as in the case of the Montagne de la Cour project (1897–98), which necessitated the demolition of the low-income Saint-Roch quartier. Another dramatic example is Poelaert's beaux arts Palais de Justice (1866–83), which supplanted a substantial fraction of the low-income Marolles quartier.

During this French period, Brussels became an unrecognizable reflection of its prerevolutionary self. Belgian nationalism with a French flavor served as a veritable bulldozer not only of the ground plan and old buildings but also of social structures competing for space in the Brussels Pentagon and surrounding areas. The nineteenth-century conception of nationalism as a search for a national identity through the crafting of new imagery combined with the drive to beautify and improve living conditions with the help of new technologies, and helped transform Brussels into a caricature of Paris. Brussels was becoming a nineteenth-century French-style capital exhibiting the urban sensibilities of Haussmann's Paris. It was a clear example of the projection of an ideology on the built environment: the desire to do away with the past in the dawn of an age of steam and iron, to replace the stagnant artisanat with the wonder of industrial production, to clear unhealthy and ill-lit neighborhoods and refuse-filled alleys with ordered, clean streets and two-story townhouses. Possibly, too, these transformations can be viewed as a primitive form of social engineering by means of city embellishment and modernization. Whatever the motive, the result was that the city managers eradicated much of the old Pentagon city and displaced

38. H. Moke, letter, in *La Belgique monumentale, historique et pittoresque*, book 1 (Brussels: 1844).

much of the original population, which was of Flemish culture and low income.

The Emergence of an International and Supranational Brussels

In the latter part of the twentieth century Brussels and the quartiers Léopold and Nord-Est have been changing at a dramatic pace. Nineteenth-century urban planning achievements set the basic frame for what has been taking place. Following World War II, the city continued to expand, especially to the south and the southeast, and the transportation infrastructure was enhanced with the completion of the gare Central, the downtown train station, in 1952. Extensive nationalist-period changes have given the city a generally uniform visual character and only excursions into the older parts of the city—inside the tenth-century fortifications—give a glimpse of what the city looked like even only one hundred years earlier. The stucco or stone neoclassical and eclectic façades of the townhouses, as well as the arrangement of these townhouses in groupings imitative of classicizing palaces, had become the norm at the expense of the traditional Flemish Brussels architecture.

It is suggested here that this so-called French-period landscape should be regarded as continuing until the 1960s when the majority Flemish constituency succeeded in achieving elements of its decades-old agenda for equal representation in government, the public sector, cultural institutions, and especially education. Brussels was again the symbolic battleground between the Walloon and Flemish cultural and political realms.[39] The Flemish community asserted its right to a city which historically, culturally, and geographically lay in Flanders, while the Walloon community sought to preserve the political and cultural establishment it had founded in Brussels since independence. In addition to overturning what they perceived as French cultural imperialism, Flemish nationalists were equally concerned with limiting the "French oil-slick" of the Brussels metropolitan area, which was threatening traditional Flemish Brabant with absorption and assimilation.

---

39. Alexander B. Murphy, *The Regional Dynamics of Language Differentiation in Belgium: A Study in Cultural-Political Geography* (Chicago: University of Chicago Geography Research Paper no. 227, 1988); and "Ethno-nationalism and the Social Construction of Space," paper, 1990 Toronto Meeting of the *Association of American Geographers;* K. McRae, *Conflict and Compromise in Multilingual Societies: Belgium* (Waterloo, Ontario: Wilfried Laurier University Press, 1986).

Two factors diluted the impact of this Flemish-Walloon cultural and political struggle for Brussels and ushered in what is proposed here as the "international" period and the ensuing "international" urban landscape of east-central Brussels. First there was the convergence on Brussels of thousands of Moroccan, Tunisian, and Turkish low-income immigrants and the rapid increase of the foreign resident population which until then had consisted of Italian, Spanish, French, and Dutch immigrants. While the city's Belgian population was slowly graying, decentralizing, and moving to the outlying greener municipalities and suburbs between 1961 and 1970, the number of foreign residents multiplied by 2.5 and came to represent 16 percent of the urban population.[40] According to the 1981 census, foreigners represented a quarter of the population of the city. Commune-specific accounting would show that some neighborhoods were 90 percent foreigner-occupied.[41]

Secondly, the cultural ambiguity of Brussels and its location in the heartland of Western Europe were advantageous as bases for the locating of the executive branch of the European Communities. The original signatories of the Treaties of Rome avoided a contest between Bonn and Paris by placing the executive branch in what was perceived as symbolically neutral ground between Romance and Germanic cultural realms. After all, from the perspective of the Great Powers, the nineteenth-century raison d'être of Belgium was that of a buffer state between France and the German Confederation. What ultimately made—and perhaps continues to make—this locational choice palatable to the European partners is the provisional status of Brussels as the seat of the European Union executive and its appended institutions.

The 1960s marked significant changes in the urban landscape of Brussels along with political and demographic realignments. The construction of high-rise buildings in the international style introduced new aesthetic and functional elements in Brussels. Large building volumes, the use of industrial materials (steel, chrome, reinforced concrete, glass), and most important, the market-deterministic and general carelessness with which these built forms were grafted into nineteenth-century Brussels, contributed to the emergence of an

---

40. R. André, "Evolution régionale de la population étrangère de Belgique d'un recensement à l'autre 1947–1981," in Jacques Lemaire, ed., *Immigrés: Qui dit non à qui?* (Brussels: Editions de l'Université de Bruxelles, 1987).

41. J.-P. Grimmeau, "Caractéristiques fondamentales de l'éspace bruxellois," *Revue belge de géographie* 4 (1985): 216.

especially unflattering aesthetic that Belgian and foreign planners alike refer to as "Bruxellisation."

What lay at the base of this laissez-faire approach to city management were the lack of detailed and enforceable zoning regulations, the desire of the local political milieu to cater to national political parties and national interest as opposed to the interests of the city and its inhabitants, and lastly, the lack of critical mass, and perhaps financial means, among grassroots activist organizations interested in urban conservation. Moreover, favorable fiscal legislation for the siting of multinational corporate headquarters in the city has, since that period, both encouraged corporate investment in real estate and business management activities, and created a nearly irresistible vogue for a "placeless" corporate aesthetic. The emergence of Brussels as the provisional administrative center of a federating European Union of states has made it even more of a magnet for organizations, corporations, and internationally minded, export-oriented entrepreneurs. Brussels and the quartier Européen-Léopold in particular are increasingly becoming the preferred location for institutions with a Europe-wide expression, and of firms hoping to benefit from the centralization of information concerning European integration.

The great number and international origin of institutional newcomers suggest that Brussels as "Europe's capital" has an appeal which extends beyond the boundaries of the European Union, since the reshaping of Europe as a trading and political block clearly has implications for the world economy and international security. The political weight of the European Union can increase internationally only with the accession of new members and the cooption of the Russian Federation. Nearly ten years following the drafting of the white paper on the Single European Market that triggered new interest in European integration, the stakes are immensely higher: will a union of fifteen members be able to make the necessary hard choices about restructuring decision making and the institutions in the Intergovernmental Conference of 1996–97? Will diversity of opinion, national interest, capability and resources, and political will among the partners ultimately cause paralysis? Will external crises, such as the one in Yugoslavia, create fatal divisions in the fragile consensus over common defense and foreign policies? Yet if the European partners manage the transformation smoothly, the rewards are probably immense. The functional centrality of Brussels and especially the quartier Européen-Léopold will translate to the attraction of the top service-sector firms, as well as governmental and nongovernmental organizations. Land rents will skyrocket. The city will harvest significant returns directly in the form of property taxes, and indirectly in the form of manifold increases in consumer spending in commercial establishments. Although the rewards for Brussels as a city will be unquestionably of historical proportions, not all constituencies among the inhabitants rejoice over the intensifying internationalization.

Before we speculate, however, about what may be, it is important to examine what in the European integration exercise has already had an effect on Brussels. The European integration dividend did not have its full impact on the cityscape until 1987, when the political will was found among the leadership of the then twelve member-states to move with a fast-paced, radical program of integration—the "1992" campaign. The implementation of the 1992 ideal did not appear to affect the rivalry between the Flemish and Walloon constituencies competing at the local, Regional, and federal levels, since both saw the emergence of Brussels as a world-class urban player as an advantage to their discreet nationalist goals. The process of internationalization, however, is changing the city in ways incompatible with the cultural character either group originally envisioned or desired.

In certain respects, the process of internationalization appears to have spun out of the control of national players, as significant building projects originate in the boardrooms of international financial players, speculative urban developers, and marketing firms. While Brussels was traditionally perceived as a market lacuna between the high-priced property markets of London, Paris, and Amsterdam, the 1985 signing of the Single European Act has jolted the Brussels real-estate market out of its slumber. It appears that the new "Europeanism" is serving as an ideological steamroller in the hands of industrial investors and real-estate speculators. "Europeanism" has often been offered as justification by developers and investors, as well as notably the Belgian and local governments, for the rezoning of certain neighborhoods and the procurement of permits for the replacement of old buildings with new office blocks and condominiums.

The commodification of Brussels's administrative built landscape and increased internationalization of the city, then, have caused shifts—not all to the good—in the morphology of residential, professional, and public space. Today we observe that the efforts of private, state, and city historical preservation agencies to conserve the existing building types, the city's townhouse ambiance, and architectural heritage are continually being challenged by the private business sector. City planners and authorities at the local and Regional levels resist pressures by the private sector to allow the emergence of mono-functional neighborhoods of offices and commercial establishments, first in fear of reducing the size of the tax-paying population base; second, in concern over losing federal grants and benefits which are calculated on the basis of population size; and perhaps most impor-tant, in alarm over losing control of costs for the provision of reason-able social housing in the city.[42]

From the interwar period onward and at an ever accelerating pace, there has been a shift away from rowhouse Flemish, Italo-Flemish, Spanish, Palladian, neoclassical, Flemish revival, art deco, and art nouveau building types, which had been standard in the city up to World War I, toward larger-scale, modern urban and suburban apartment and office blocks, single- and two-family suburban houses, and renovated old townhouses. Significantly, these transformations included the reapportioning of interior spaces: in the case of residen-tial space, the subdivision of single-family townhouses into two or more apartments with modern amenities challenges the established concept of what has constituted standard or adequate family space in Brussels.

Yet there is a very strong correlation between the ethnic composi-tion of certain communes and private investment-driven gentrifica-tion. Christian Kesteloot very appropriately states that there is an emerging sociospatial polarization in Brussels, where ethnicity, age, family size, income, and occupation increasingly define social and spatial boundaries between the affluent and the needy. He notes that these populations "are subject to very strong forces of marginaliza-tion, which were unleashed by the economic crisis on the one hand, and by urban revival and the local effects of a flexible service econ-omy and interurban competition on the other hand."[43] His study builds on his earlier work with Walter De Lannoy on residential dif-ferentiation and segregation as a by-product of land markets.[44] Kesteloot and De Lannoy note the persistence of patterns of sociospa-tial separation at least as early as 1970, when the first statistical sur-

---

42. Cabinet du Ministre-Président, *Forum: L'Immobilier à Bruxelles* (Brussels: Executif de la Région Bruxelles-Capitale, 1990).

43. Christian Kesteloot, "Three Levels of Socio-Spatial Polarization in Brussels," *Built Environment* 20, no. 3 (1994): 212.

44. Walter De Lannoy and Christian Kesteloot, "Differenciation résidentielle et pro-cessus de ségrégation," *La cité Belge d'aujourd'hui: Quel devenir?* In the 39th annual special issue of *Bulletin trimestriel du Crédit Communal de Belgique*, no. 154 (October 1985).

vey of Brussels on the basis of 567 census tracts was conducted.[45] In brief, they observed a strong polarization of the Brussels population on the basis of employment. Moreover, they detected sociospatial polarization among groups of foreigners: nationals from the countries of the Maghreb, Turkey, Italy, and Greece tended to congregate in or close to the Pentagon, where the housing stock is often in relative disrepair, whereas nationals from the affluent states of the European Community and Americans tended to congregate in the more affluent communes to the east and southeast. The correlation of the patterns of settlement based on employment and ethnicity is strong: North Africans are more likely to live in working-class neighborhoods, while Northwestern Europeans are usually living in white-collar neighborhoods.[46]

Today Moroccan, Tunisian, and Turkish communities continue to be segregated in the northern and western parts of the city, which were once home to most of the city's industry and which contain some of the oldest and lowest-quality housing stock in the city. In these architecturally frozen neighborhoods, the residents recreate the landscapes they left behind if only in custom, dress, diet, language, and worship. On a Sunday morning promenade at the food bazaar around the gare du Midi one has the impression of being transported to Marrakesh or Tunis as one walks among turbaned, jellaba-clothed men and haik-swathed women bargaining in Arabic and Turkish with cheese, olive, and spice sellers. The sobering side of this exotic landscape building is that investors—especially the large corporate and institutional investors—systematically avoid communes with large non-European immigrant populations. What started as an informal process of segregating non-Europeans in some of the deteriorated nineteenth-century extensions of the city during the 1950s and 1960s, has become a well-established practice among real-estate professionals advising Belgian and foreign investors alike. In my personal survey, I heard many times cautionary tales about investing in such neighborhoods, where returns are modest, turnover of renters and lessors frequent, and the trajectory of land values rather flat. Such campaigning, although not necessarily stemming from malice, has tremendous implications for the development of these neighborhoods. The current land marketing climate essentially condemns these populations to modest social and physical plant improvements

---

45. In the 1980s the number of census tracts *(secteurs statistiques)* was increased to 722.

46. De Lannoy and Kesteloot, pp. 139–49.

generated by the modest capital resources of the inhabitants and the beneficence of the Belgian state and the Brussels Regional government. As Kesteloot alludes, this resembles ghettoization. The full social implications of this process have yet to be fully understood or felt.

The process of social polarization also extends to certain Belgian middle- and working-class neighborhoods in the northern and western parts of the agglomeration. Perceived as distant from the strategic real-estate markets and touristic ancient city and exhibiting smog-covered façades, they have not captured the imagination of important local or international investors and speculators. In a city where housing prices are rising rapidly out of reach for most of the working- and middle-class population, this temporary respite may be a blessing. In real terms, however, these populations have less and less access to a city and services that are increasingly geared toward high value-added tertiary and quaternary activities that are initiated abroad. Interestingly, however, the apparently similar plights of the non-European and the Belgian working-class neighborhoods have not given rise to a united political front in the city. Such divisions make the agency of the government/private-sector block nearly irresistible.

Frontier neighborhoods are as indicative of the revolutionary changes occurring in Brussels as neighborhoods excluded from gentrification or development. These are the areas of the city contiguous to the foci of international interest: areas in the eastern and southeastern sectors of the city, close to the administrative quarter of the European Union and to the axes favored by high value-added industries and services. These appear irrevocably on course for gentrification and redevelopment.

Until the launching of the "1992" campaign in 1987, the green spaces of the art nouveau squares Marguerite, Ambiorix, and Marie-Louise, only two hundred meters from the European Commission headquarters, were the frontier between the administrative quarter and the quiescent, largely Turkish neighborhood to the north. Today, elderly Turkish men share the squares with lunching Commission officials as gentrification is making major inroads in the frontiers of the previously Turkish neighborhood, claiming dozens of townhouses leading north and away from the squares (rues des Patriotes, de la Brabançonne, Pavis, boulevard Clovis, and rue Waterloo Wilson). A similar dynamic exists near the place Jourdan, to the west of the rue du Midi, along the avenue de Stalingrad, along the streets Haute and Blaes in the southern section of the most ancient part of the city (the environs of the Grand Place), and in the suburbanized rural communities such as La Hulpe, southeast of the Brussels agglomeration, which hosts the largest IBM administrative facility in continental Europe and SWIFT (Society for Worldwide Interbank Financial Telecommunications).

At the same time "modern" neighborhoods arise in traditional residential enclaves of the Belgian and western European social establishment. These *beaux quartiers* are found at Uccle (avenues Molière and Brugmann), Ixelles (Jardin du Roi, Etangs d'Ixelles, place Leemans), Etterbeek (La Chasse), Woluwe-Saint-Lambert (avenue de Broqueville and boulevard de la Woluwe) and Woluwe-Saint-Pierre (avenue de Tervueren, Val Duchesse), and highly developed service industry areas, such as the quartier Léopold, the avenue Louise, and the boulevards du Souverain, de la Hulpe, and Léopold III. These areas were first affected by the growth of the property market.

On the one hand, these high-market neighborhoods provided top-quality period houses as residences. On the other hand, they satisfied the need for addresses of prestige for businesses, relative proximity to frequent business partners and governmental institutions (the federal Belgian Cité Administrative, the European Commission and Parliament complexes), and adequate accommodations for the executive operations of multinational corporations, representative trade and industrial organizations, lobbying bodies, chambers of commerce, commercial and investment banks and insurance corporations, and foreign consulting and law firms. Professional space for political institutions, corporations, and organizations is found increasingly in block-sized office towers and in low-rise buildings in modern or postmodern styles.

The urban landscape of Brussels bears the mark of the recent constitutional reform which promulgated a national federal system and created largely separate regions: Flanders, Wallonia, and the Brussels-Capital Region. These three Regional administrative entities require infrastructure and space in the city of Brussels separate from that of the Belgian federal government. For example, a large complex is being built on the boulevard Léopold II to accommodate the Regional representation of Flanders; this construction is serving to initiate the redevelopment of neighborhoods immediately northwest of the Pentagon.

Most important for our purposes, however, are the marks Brussels bears from its role as seat of the EU executive branch. Congestion around the ever-growing complex of the European Commission is of such proportions that in early 1989 local structural adjustments had to be made to the recent highway network. This institutional growth has been responsible for the recent extraordinary proliferation of office buildings in Brussels. Again, in the quartier Européen-Léopold, many of the mansions built by the aristocracy and prosperous bourgeoisie during the middle of the nineteenth century have been either taken over by corporations or demolished to make way for large office blocks. In 1974, long before the urban boom, enough offices had already been built to replace a large fraction of the quartier's 1930 population.[47]

The accepted nomenclature of urban change in Brussels in this last period encompasses "growth," "redevelopment," "gentrification," and "evolution." Like urban preservation and conservation, urban change is an ideological movement that is viewed differently by different inhabitants of the Brussels landscape. A windfall investment for the Swedish investor, "a find" of a townhouse for the executive newcomer of the European Commission, or an improvement in the taxable base of the city for the city budget officer—all positive developments from a certain point of view—mean unaffordable housing for the elderly and for the working and middle classes, economic segregation of ethnic minorities, increased air and noise pollution, reduction of open and green spaces in and around the city, and pressure on the transportation infrastructure resulting in time-costs for all. The situation is aggravated by the fact that federal regulations have fitted Brussels with an iron collar limiting its physical growth into the surrounding Flemish Brabant. The competition for space in the city will therefore be all the more intense.

Conclusion

Brussels was founded in the tenth or eleventh century as a modest fort and hamlet on the fringes of Roman Europe. Its subsequent steady assent from a hamlet to a city of some Regional economic and political significance in the thirteenth century, and a center for the production and trade of magnificent tapestries and luxury goods by the fourteenth century, was due to its advantageous location along East-West trading routes. With the ascent of Charles V to the Habsburg throne Brussels was suddenly propelled to the position of imperial capital and to the center of dynastic rivalries, with dramatic consequences for the political and material structure of the city. Suc-

---

47. Smolar-Meynart and Stengers.

cessive domination by Spaniards, Austrians, French, and Dutch has left its imprint on the city.

By the time Belgium gained its independence from Holland in 1830, the city was already in the grip of industrialization, nationalism, and the democratic revolutions which swept nineteenth-century Europe. The domination of the polity and the economy by a French-culture elite and the preeminence of French urbanistic and architectural sensibilities turned the once Flemish city into a French-style capital, at the cost of demolishing the material legacy of Flemish culture.

Political, economic, and aesthetic internationalization have changed the city again. Factors such as the Belgian export-oriented economy, the necessity of attracting foreign direct investment and international corporations in order to keep the national economy globally competitive, and finally the siting of international administrative bodies like the European Union in the city have diluted the national capital character of the city and have given rise to a new conception of the European executive downtown: one which focuses on high administrative functions relating to the process of Regional economic and political integration in Europe. This specialized executive downtown is uniquely devoid of major retail and entertainment activities we usually identify with traditional CBDs, and may warrant the new label "central executive district," or CED.

The comparison of Brussels with Lichtenberger's Vienna as "eurometropolis," or political interface between the West and the newly liberated Central Europe, is useful and unavoidable. Although it attained world-class status before the First World War as a result of its imperial legacy, Vienna, Lichtenberger claims, is poised for a new era of greatness and a new founders' period. Vienna's new role as "eurometropolis" in "Paneurope" is defined, as is Brussels's, by the reinterpretation of statehood, sovereignty, and economy by European integration.[48] Brussels is a key city in the European Commission's conception of how European integration should be managed, and a major node in the European executive heartland which extends from southeastern England to Northern Italy. However, in light of the

---

48. Whereas Vienna layoutside the megalopolitan region conceived by the Commission as the urban-economic cortex of the European Community before the liberation of the Eastern Block, Lichtenberger vies to redefine the European megalopolitan region to include Vienna as the key entrepôt and broker between Western and Central Europe and the Balkans. Elisabeth Lichtenberger, *Vienna: Bridge Between Cultures* (London: Belhaven Press, 1993), pp. 181–95.

Union's commitment to equitable Regional development within the confines of the Single European Act, and the need to satisfy a growing number of political and ethnic constituencies, the question remains how much more central to the European integration process can Brussels become before political forces erode its position in favor of a more Regionally diffuse European administration. Moreover, how well do the Belgian government's and private sector's urban conceptions for the seat of the European executive branch fit the needs of the city's inhabitants, its strategic planning for the twenty-first century, and the image Brussels wants to project to the world? If the European integration enterprise stands for pluralism and the democratic process, how appropriately does the urban strategy that is giving rise to Europe's central executive district fit these ideals?

4

The Planning Aspect

B RUSSELS IS an atypical Western European city because of the historical incidents that made it a capital city, because of its peculiar urban regulatory tradition—or lack thereof—and because of its recurrent flirtation with American-style urban development practices and planning. Each of these factors deserves separate examination.

Historical Forces

Some of the points made in chapter 3 can be recast in reference to the role of the free market in Brussels. Specifically, the role of the private sector in planning decisions has been basic to Brussels urbanism. Since Brussels became the capital of the Kingdom of Belgium in 1830, finance capital and the private sector, as represented by the landed aristocracy and prosperous bourgeoisie, have been key participants in the modernization and expansion of the city.

The task of turning the modestly sized ancient pentagonal city into an industrial-era capital was great. The resources of the German-born prince who became Leopold I of Belgium were certainly limited. Yet a historic opportunity appeared for the holders of surplus capital—nationalists and foreigners—to invest in the modernization and expansion of this new European capital. It was a private concern that launched the opening of the avenue Louise, and a different private concern, the Société civile pour l'agrandissement et l'embellissement de la capitale de la Belgique, which undertook the planning and construction of the quartier Léopold. Its executive offices and stockholders were attached in one way or another to the French Société Générale holding company, the great financial concern which today continues to influence the quartier.

The quartier Léopold united in its form the state of the art in urban financing and urban planning. Conceived by private investors after the pentagonal fortifications had been demolished, the plan of the new quartier was to follow the newly fashionable ground plan *en damier* (checkerboard), which facilitated the parceling of standard-sized street blocks and the marketing of lots large enough to accommodate the construction of mansions and monumental public buildings. This orthogonal extension of the city adopted the orientation of the rectilinear street blocks of the rue Ducale, which separated the baroque parc de Bruxelles from the avenue that had replaced the demolished pentagonal fortifications. Thus, the mid-nineteenth-century quartier Léopold was to provide a harmonious morphological transition between the "ancient" city and the "modern" one (fig. 12).

The wishes of the royal government and Palace for great expenditures outside the pentagonal city were contrary to those of the liberal city council of Brussels, which feared losing wealthy inhabitants to the periphery where it could not impose taxes, or control city institutions. The royal government prevailed and the quartier Léopold was launched and eventually annexed by the city in 1853. There are many examples of this sort underlining the significance of the private sector as implicit contractor to the Palace: the extension of the rue de la Loi, the launching of the quartiers Nord-Est and Koekelberg, and the demolition and "Haussmannian" transformation of the quartier Notre-Dame-aux-Neiges in 1877. Whereas the promoters and beneficiaries of changes in the quartiers Léopold and Nord-Est belonged to the most prosperous stratum of Belgian society, the promoters and the beneficiaries in the case of the quartier Notre-Dame-aux-Neiges were principally smaller investors.

> In the beginning, the buyers [in the quartier Notre-Dame-aux-Neiges] were individual private persons who generally purchased a single lot to construct a house, either for the purpose of renting it to others, or to occupy it themselves. Occasionally small property owners invested in two or three lots, but this was a rare occurrence. In a second phase, things started happening fast, and the buyers were professionals who favored the location of the quartier. They were businessmen and architects who would purchase a number of contiguous lots to build houses which they hoped to sell for profit.[1]

---

1. Translated from C. Jacqmain, "Les operations de la société anonyme du quartier Notre-Dame-aux-Neiges (1874–1888)" (Thesis, Université Libre de Bruxelles, 1976), 1:93.

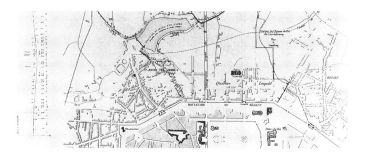

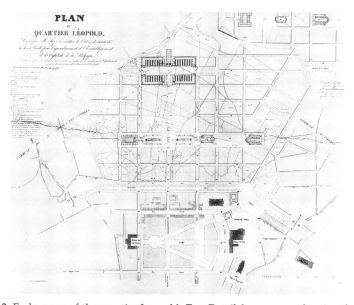

Fig. 12. Early maps of the quartier Leopold. *Top:* Detail from a map showing the projected extension of the rue de la Loi (1852), also showing the first buildings constructed in the quartier Leopold. *Bottom:* Early map of the quartier Léopold (1837), commissioned by the Société civile pour l'agrandissement et l embellissement de la capitale de la Belgique.

These practices are certainly not native or unique to Brussels but bear some significance for our understanding of who "owns" Brussels today, and who is trying to buy a share. Since the completion of the quartier Nord-Est around 1915, the owners and occupants of the quartier Européen-Léopold have been changing from Belgian individual owner-occupiers and small *rentiers*, or real-estate investors, to Belgian, foreign, and international corporations and institutional clients. (I concern myself here with the owners; the participants are discussed in chapter 6.) These changes in market control signify a sharp departure from the historical market environment which gave rise to the quartier, and therefore will produce a very different urbanism.

If we consider the nature of the quartier as a market in 1915 and 1992 we find staggering differences. In keeping with the two principles that guide urban land prices and urban land use—"for any parcel, land price is what the highest bidder is willing to pay," and "the more expensive the land, the greater are the incentives to economize on its use through more intense, or higher density usage"—the new historical conditions of European integration and the building of Europe in Brussels bring in economic actors who are willing and able to bid up the price of land in our quartier and to jettison the small owners. The need to economize on this precious EC-anointed land means the replacement of the low-density urban fabric with a denser, taller, less green, more crowded urban fabric, designed for the tasks of European management.

The key amenity sought by land consumers of the quartier is access to the European institutions and the political individuals who work within them. By locating offices on the widely known streets and avenues of Belliard, Loi, Joseph II, Cortenberg, Auderghem, Arlon, and Trèves, for example, one purchases legitimacy if not with the EC then with the rest of the business world. The question then becomes what the perspective buyers of space want to buy. Is it fin-de-siècle townhouses like the few surviving on the rue Toulouse and the rue De Pascale, baroque revival mansions like the remaining two flanking the Church of St. Joseph at the square Frère Orban, specially priced modernist relics from the first phase of expansion of the office sector in the quartier, or a gleaming postmodern building in bronze, porphyry, and glass? The market is largely unencumbered by structural constraints in determining what stays and what goes. This situation is, however, slowly changing.

The City of Paper Regulations

Barring the frequent complaints of real-estate development firms concerning red tape and the astonishing slowness with which building permits are considered and approved, Brussels appears to be a laissez-faire, or "laissez-bâtir" paradise. Cynics may claim that money talks and that political influence is the only factor of importance in Brussels urban planning, yet regulations abound.

While Sweden has had a legislative framework for land management and urban planning since 1874, the Netherlands since 1901, Germany since 1904, and Britain since 1909, the first serious attempt to introduce urban planning in Belgium took the form of the so-called organic law on land-use management and urban planning ratified 29 March 1962.[2] Its impact on urban planning remained negligible because it essentially codified the procedures that would need to be followed by the appropriate authorities in making decisions. Amazingly, the 1962 law did not set down guidelines, standards, or short- or long-term goals for cities and villages and their planning authorities where these existed. Bowing to political and social pressures during a time of national agitation over budding federalization, the law attempted to integrate all the forces in society in the land management process. The government clearly stated its motive:

> We do not know what needs to be done, what types of management to encourage so that the public reaps the greatest benefit, but when it will come to taking a decision, we want to foster dialog among all the social forces represented in the country, in the Regions, and in the municipalities.[3]

The vagueness of this law did not encourage serious planning for cities and villages but instead encouraged laissez-faire practices.

Since 1962 and despite the settling of the federalization question, there is neither a national nor a Regional *plan de l'aménagement du territoire* (land-use plan). Even now that regions are legally fortified, any suggestion that the national government should control planning would raise suspicions among the regions.

The royal decree of 28 December 1972 instituting the *plan de secteur* called itself "the first concrete step taken toward producing a planning vision together with certain standards for the allocation of land to different uses."[4] The decree set forth a zoning classification which included four subclasses of residential zones (from pure resi-

---

2. Réné Schoonbrodt, *Essai sur la destruction des villes et des campagnes* (Brussels: Pierre Mardaga, Série Architecture et Recherche, 1987), pp. 135–37.

3. Ibid., p. 136. Schoonbrodt quotes from a government source without citation.

4. A formal, countrywide definition of *secteur* escapes even the people who work in the Brussels metropolitan area planning office. In the case of Brussels, *secteur* encompasses the nineteen municipalities of the Brussels metropolitan area (Agglomeration de Bruxelles).

dential to a variety of mixed-use space), industrial/commercial/
warehousing zones, office zones, zones for large-scale commercial activities, zones for social services such as schools, sports, and cultural
facilities, zones for public and administrative services at the local,
Regional, national, and international levels, zones of green space in
quartiers, specifically green space associated with buildings (i.e.,
gardens in the interior of street blocks), zones of public green space
and vacant land, zones for cemeteries, zones for roads, railroads, and
waterworks. This modernist vision encouraged the separation of land
uses and the formation of pure zones. Residential zones opened to a
great variety of other land uses, including the building of office
blocks, appeared to be at a disadvantage.

The Brussels College of Mayors and Aldermen[5] ratified a *plan de
secteur* on 28 November 1979 (fig. 13). Owing to pressure from
groups such as the Atelier de Recherche et d'Action Urbaines
(ARAU), the plan was amended to provide protection for zones and
sites of cultural, historic, or aesthetic interest. Under this plan, such
zones are intended to preserve all of the quartier Nord-Est, the
square Frère Orban, including all the buildings surrounding it, the
streets Pascale and Toulouse in the quartier Léopold, the parc Léopold, including the rue Vautier, all of the square de Meeûs, all of the
square de Luxembourg, including the Luxembourg train station
designed in 1855 by the architect Paul Santenoy, and a small portion
of the rue Joseph II at the intersection of rue Philippe le Bon.[6]

Thirteen years later many of the structures that these special zones
were intended to protect are still standing, some splendidly renovated, such as the mid-nineteenth-century baroque revival mansions
on either side of the Church of St. Joseph on the square Frère Orban.
Splendid period buildings across from them have however been
replaced by mid-rise office buildings. The same has happened to the
buildings on the south-southwestern side of the square de Meeûs. At
the southern boundary of the quartier Nord-Est, facing the Berlaymont is situated the International Press Center, home to the second
largest press group in the world after that of Washington, DC.
Construction for the Centre International du Congrés (i.e., the ex-

---

5. The Collège des Bourgemestre et Echevins runs the Brussels metropolitan area. It
brings together the leadership of the nineteen municipalities comprising the metropolitan area.

6. Secretariat d'état à la région Bruxelloise: Administration de l'urbanisme et de
l'aménagement du territoire, *Plan de secteur de l'agglomeration Bruxelloise: Situation existante de droit et de fait* (Brussels, 1979), map 2.

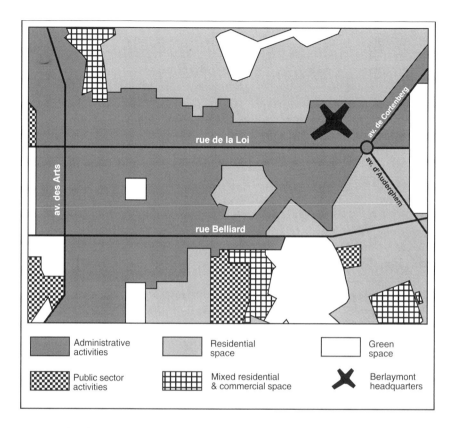

rue de la Loi

av. de Corterberg

av. des Arts

av. d'Auderghem

rue Belliard

| | Administrative activities | | Residential space | | Green space |
| | Public sector activities | | Mixed residential & commercial space | | Berlaymont headquarters |

Fig. 13. The *Plan de Secteur* of 1979

pected seat of the European Parliament) has closed off the north side of rue Vautier. Ten street blocks in the quartier Européen-Léopold which were zoned exclusively for residential use or mixed residential and business uses have now been almost entirely absorbed by EC-related building projects.

Indeed, the *plan de secteur* was conceived and functioned as an impressionistic vision of Brussels. It was never intended as a blueprint, which is a fact that neither pressure groups nor the public understood at the time. To the disappointment of some, it has been the only urban planning document yet to emerge from democratic cooperation among government agencies, planning think tanks, the business community, and committees representing the residents of Brussels. While *plans de secteur* are a step in the right direction after a long planning vacuum, they are seen by constituencies favoring strict regulation of planning to fall far short of protecting housing and the mixicity of land uses in cities.[7]

While provisions exist for metropolitan area-wide urban management plans, so-called *plans généraux d'aménagement* are rarely the basis for urban transformation. As a result this entire level of planning is at this point irrelevant to real change, just as is national and Regional-level planning.

Real planning authority rests at the microscale: in the design and approval of the *plans particuliers d'aménagement* (PPA), or "specific management plans," which are parceling and building permits. The PPAs, which need to be approved by royal decree, make piecemeal changes in the general vision of the *plan de secteur*. PPAs turn Brussels urbanism into a palimpsest where, theoretically, specific needs of the city or market are addressed without disrupting the environs of the affected area. Each PPA is entirely independent of other PPAs and may have a variety of purposes. While a PPA usually controls a few street blocks of area, there are no formal limitations to its areal extent. For example, the Brussels municipality PPA no. 41–31/32 approved on 17 January, 1964 involved the expropriation and leveling of all the townhouses in the single street block bounded by rue de la Loi and the avenues de Cortenberg and de la Joyeuse Entrée, and the construction of the "Triangle" building housing the EU's translation service. By contrast, the Woluwe-Saint-Pierre PPA nos. 1A and 1B, approved on 1 February 1956, involved transformations in fifteen street blocks.[8]

The royal decree of 5 November 1979 set down rules for the publication of suggested PPA, parceling, and building permits. Specifically, the decree required the timely placement of posters detailing the changes proposed in a PPA or parceling or building permit in the affected area and its environs. It furthermore gave the public access to the technical information involved (one day per week for either two or four weeks depending on the nature of the project). Additionally, it allowed for the creation of special consideration commissions (*commissions de concertation*) responsible for identifying flaws in the proposals and lobbying in favor of local concerns. Its members are drawn from the executive branch of the region, the municipal planning authority, the Council of Mayors, the Metropolitan Council, and the administration of the Agency for the Regional Development of Brussels (Société de développement régional de

---

7. Schoonbrodt, p. 148.
8. Secretariat d'état à la région Bruxelloise, pp. 13, 34, and map 2.

Bruxelles). Although the royal decree allows for additional members, the commissions are heavily packed with government officials. In theory, the views of private industry and residents are filtered through these commissions. (In chapter 6, I will explore the extent to which a review process dominated by politicians may fail to reflect the views of all parties concerned.)

Predating these recent efforts to introduce game rules into urban planning, a regulation relating to the construction and aesthetics of buildings has existed since before Belgian independence. Starting with the 1789 regulations concerning structures (*règlement sur les bâtisses*), individual municipal authorities have slowly built a rather complex code which addresses technical aspects of building. Since neither national nor Regional authorities have taken the initiative— or had the political capability—to institute national or Regional standards for building, the regulations differ significantly from city to city. They do not as a rule address any aspects of urban planning other than the general height of buildings in relation to the width of the particular street, and the aesthetics of façades.[9] The main body of the code pertains to the safety of construction, protection against fire and flood, the proper connection between the sewage and water systems of the buildings and that of the city, and the designation of basic building amenities.[10]

What does all this mean for the market? The existence of four active layers of planning (*plans de secteur*, PPAs, parceling, and building permits) ostensibly entails considerable transaction costs for market participants. In a conference on the Brussels real-estate market organized by the president of the Brussels-Capital Region on 15– 16 March 1990, the participants representing substantial construction, real-estate development, and management concerns, such as Jones Lang Wootton, Jacques Delens, and Tractebel, complained bitterly about what they called "inexcusable delays in the processing and approval of parceling and building permits."[11] Indeed, the review process can take more than a year. In the overheated market of the late 1980s, any unnecessary delay represented significant opportunity costs. In contrast to individuals in the private sector, there

---

9. Marc Eloy, *Influence de la législation sur les façades bruxelloises* (Brussels: Commission Française de la Culture de l'agglomeration de Bruxelles, 1985), p. 3.

10. Ville de Bruxelles, *Règlement sur les bâtisses* (Bruxelles, 1981). Volume includes a number of loose-leaf appendixes bringing the code up to date.

11. Cabinet du Ministre-Président Charles Picqué, "L'immobilier à Bruxelles: Forum," address at the Hôtel Palace, Brussels, Conference, 15–16 March 1990.

naturally exist critics who would like to see significantly more regulation of the urban planning process and the containment of certain
market tendencies—namely the explosion of the office sector in the
quartier Européen-Léopold. A long review process may be desirable
if it produces thoughtful planning suggestions. The worst possible
scenario, of course, is that of a lengthy review process that does *not*
produce thoughtful suggestions or amendments to proposed projects.
To determine whether this is the case, we have only to look at the
end result: the built environment of the quartier.

The various layers of planning regulations do not in reality result
in significant obstruction of market forces. The palimpsest method of
urban planning in Brussels, and especially the use of PPAs, makes
for occasional, point-specific transformations of the landscape. Its sup-
porters claim that in this manner only small parts of the landscape
are affected. Its critics claim that it has a domino effect around the
city: Concentrations of PPAs in certain areas of the city weaken the
existing built fabric. This appears to be true in the case of the
quartier Européen-Léopold. Real-estate developers have frequently
argued that once the majority of the nineteenth-century buildings in
a street block have been removed, then all the others may be
removed as well, since the cultural/landscape amenity is rendered
insignificant (fig. 14).[12] By the same token, when a given area is
dominated by modern street blocks, then the one or two remaining
traditional ones may just as well be absorbed.

While planning regulations may not serve as significant market
obstacles today, there are indications that Brussels may adopt more
stringent planning parameters in the future. The shift of public opinion toward support for stricter urban planning and the support of the
housing sector, and the shift of power in Brussels and other cities
from national to Regional and local politicians as a result of federalization, suggest a coming assault on laissez-faire urbanism. Given
the planning status quo, however, we can safely maintain that the

12. Such a controversy involving a small number of neoclassical townhouses on rue
Guimard reached the press after the intervention of ARAU and IEB. More important,
the initial project for the Centre International du Congrés included the demolition or
the moving of the façade of the Luxembourg train station from the place de Luxembourg to some undisclosed location. The neoclassical façade is one of the finest works by
the architect Gustave Jean Jacques Saintenoy. Because of the public outcry, the façade
will be preserved. Jean-Claude Vantroyen, "La façade de Gustave Saintenoy, un détail?" *Le Soir* (17 May 1990).

Fig. 14. Old and new buildings and land uses on the southern frontier of the quartier. The newly constructed twelve-story annex to the European Parliament complex replaced modest fin-de-siècle rowhouses. The remaining ones are abandoned and in great disrepair.

quartier Européen-Léopold, as a modern CBD, is the outcome of largely unrestrained market forces.

Choosing a Course of Action:
High Density or a Balanced Mix of Land Uses?

It is misleading to take too insular a view of the market structure and trends of the quartier Européen-Léopold, or of any central business district for that matter. While a microcosm of shifting land uses and hills and dips of land values exist in the quartier, there is certainly a wider world of economic fluctuation that has a profound impact on the relatively modest area of our quartier. Brussels and its various business nodes—including the quartier Européen-Léopold—compete with other cities and business locations within them for business and administrative clients. Such cities as New York, London, Paris, and Tokyo have been competing to attract world-class corporations and international administrations for a long time. Brussels is a comparatively recent addition to this exclusive club and is a much smaller one in terms of office space and land prices commanded. As one may suspect, the growth of the European Communities coupled with the worldwide economic boom of the latter half of the 1980s is responsible for this development.

The Brussels-Capital Regional government is very aware of international competition for clients. In its Strategy for the Regional Management of the Administrative/[Office] Sector (Stratégie pour une géstion regionale de la fonction administrative), the Regional executive lays out the courses of action available to Brussels:

a. following the model of American cities
—where the development of the office sector in the centers of cities takes the form of skyscrapers, thus restricting the areal extent of the office district while increasing its density of occupation;
—where the development of suburban office districts relieves the pressure in the central business district.
b. calibrating the capacity of the transportation network (highways, roads, and mass transit of different kinds) in accordance with the growth of the administrative/office sector in the central business district, thus allowing for the efficient circulation of office workers between their place of residence and the central business district.

c. allowing the growth of the administrative/office sector to guide redevelopment and gentrification in the centers of cities (for example, the Docklands project in London). In this case, the administrative/office sector would be dominant in the area.

d. and finally, promoting the good balanced mix of residential, commercial, and administrative/office functions in the downtown (for example, as seen in the Paris downtown—rather than the La Défense district—and in Geneva). In this case, the traditional integration of commercial first floors (rez-de-chaussée) with residences in the upper floors will be preserved.[13]

The first three proposed courses of action reflect current national and Regional government policy with respect to Brussels and the quartier Européen-Léopold. Given that the bulk of real-estate development in the quartier is initiated, financed, and executed by the private sector, and given the favored types of construction in the quartier since the early 1960s, it is safe to assume that the private sector shares the tenets represented in the first three courses of action described above.

There is a clear preference for higher-density building in the quartier Européen-Léopold and other districts in the city that lend themselves to tertiary sector activities (the quartier Nord and avenue Louise primarily). The nineteenth-century built environment rarely allowed for buildings higher than five stories, whether low-end shophouses in the Pentagon city or mansions in the quartier Léopold. The emerging morphology of the quartier Européen-Léopold is dominated by ten-to-fifteen-story buildings. Because of this greater design-intensive efficiency in the use of space, and a decrease in working space per worker in the last two decades, the density of occupancy in the quartier Européen-Léopold has increased at least threefold.

Addressing the second of these four proposed courses of action, we see that transportation investment indeed tries to keep pace with the explosion of office tower building. Public transportation is good, though it clearly attracts less government investment than road construction. A controversial project has been proposed to build a direct rail link between Zaventem International airport and the quartier

---

13. Region de Bruxelles-Capitale, Administration de l'urbanisme et de l'aménagement du territoire, Document de synthèse, *Stratégie pour une géstion regionale de la fonction administrative* (Brussels: BRAT, March 1991), pp. i–ii.

Européen-Léopold under the avenue de la Brabançonne and the fin-de-siècle squares of Marguerite, Ambiorix, and Marie-Louise. The proposed line, which is intended to be built just below street level, would bring international business and government travelers directly to the suburban station of Schuman, a mere street block from the Berlaymont headquarters building. During the spring of 1990, the residents of the quartier Nord-Est, the portion of the quartier Européen-Léopold under which the new train line was to be built, launched a campaign to stop the project. On most windows of residential townhouses in that area one would see mounted posters in pastel green calling "Stop le tunnel!" The fear was that the new line would increase the pressure of the office sector on the residential areas, and it would create an axis of noise pollution along the middle of the largely residential, fin-de-siècle quartier Nord-Est. Additionally, the quartier is connected to the three main railroad stations of Brussels (Nord, Central, Midi) via the subway system. Line 1 of the Brussels subway runs the length of the rue de la Loi axis and includes three evenly spaced stops in the quartier: Arts-Loi, Maalbeek, Schuman. The subway is heavily used throughout the day. It stops service at midnight.

The national government reveals a clear preference for the accommodation of automobiles in Brussels rather than the expansion of mass transit. Brussels has a superb system of three largely concentric beltways that tie it to the national highway network. The highway network is rivaled only by Germany's. Quartier Européen-Léopold lies between the first and second beltways (avenue des Arts and boulevard St. Michel). It is connected to the beltways via the tunneled highways of avenue de Cortenberg and rue Belliard. As the Constitution attributes the responsibility for roads and highways to the national authorities, the road construction projects have been traditionally perceived as a first-rate pork barrel. At a time when national authority is devolving to the regions, such outlets of political influence are treasured and fought over. The result has been a marked increase in automobile circulation, air and noise pollution, and congestion in the main east-west axes of rue de la Loi and rue Belliard. The promotion of the automobile culture is evidently consistent with the city's laissez-faire philosophy of planning and the promotion of private-sector land uses.

The third course of action—trickle-down redevelopment and gentrification with office/administrative investments as the focus—has also been embraced by the national and Regional governments. The extension projects of the European Communities to the east of the parc Léopold (European Parliament), south of the Schuman round-about (Council of Ministers), and north of the place Jourdan (Centre Borchette extension—European Commission), include the construction and/or preservation of residential space. The housing created by these initiatives is relatively small because the real-estate development industry perceives investment in housing as unprofitable, especially when compared to the high returns on office development during the late 1980s. For example, the neighborhood reconstruction plan revolving around the development of the Centre International du Congrés (CIC) dedicates 17 percent of space to housing.[14] There are minor provisions for diversified commercial activity in all of the EC extension projects.

Clearly, the emergence of the quartier Européen-Léopold as a major employment area for well-remunerated Belgian and foreign administrators and business people has created a wave of redevelopment and gentrification in the areas that have not been invaded by administrative/office uses. The general perception is that "trickle-down" redevelopment produces a quantitatively and qualitatively uneven and unpredictable residential landscape. Since housing development is deemed relatively unprofitable by many real-estate development firms, it remains very much in the hands of potential owner-occupiers and small *rentiers*, or property developers. The residential areas that surround the emerging central business district to the north, east, and south are not morphologically and socioeconomically homogeneous. There are marked variations in the quality, extent, and nature of residential gentrification and real-estate investment throughout the quartier. Fashionable axes of gentrification are evident, as are enclaves that have attracted insignificant private investment to date.

The fourth course of action reflects the preference of the parties opposed to an American iconography for the quartier Européen-Léopold. Pressure groups and think tanks, such as Inter-Environnement Bruxelles (IEB), ARAU, and the Centre d'études et de recherche urbaines (ERU), as well as residents' committees and the liberal press, support the blending of residential, commercial, and administrative land uses in the quartier and throughout Brussels. The Brussels metropolitan area authorities and now the Regional government are divided on this issue. Those in favor of combining

---

14. Générale de Banque–Coopérative Ouvrière Belge, *Brochure of the Centre Internationale du Congrés* (Brussels: Générale de Banque–Coopérative Ouvrière Belge, 1987).

land uses and preserving housing in the quartier regard the fiscal contributions of tax-paying individuals and national government contributions based on the number of inhabitants residing in the city as critical to the financial health of the city. The opposition deems the collateral contributions of European Community executives, lobbyists, corporate executives, traveling businesspersons, and tourists as more than adequate sources of revenue in exchange for the modest losses in population caused by the concentration of administrative/office uses in the quartier.

Taking seriously the desires of the EC for a compact EC administrative district, those opposing the further development of the quartiers Léopold and Nord-Est as a CBD have proposed a number of alternative locations to the relocation of EC and EC-related administrative/office uses. As early as 1972–73, the Meeûs estate close to the boulevard du Souverain and viaduct of Herman Debroux had been considered a possible site for the building of the recently completed new Council of Ministers building, across from the Berlaymont headquarters building on rue de la Loi.[15]

A plan proposed by ARAU and IEB would relocate most of the EC built community to the Josaphat Station rail yards of the municipality of Schaerbeek, approximately 1.5 km north of the Schuman roundabout and well outside the nineteenth-century city. In a preliminary project long on aesthetic vision and political symbolism prepared by an architect/urban planner and a landscape architect, the two pressure groups go on the record against what they call the American-style, market-driven urbanism reproduced in the quartier Européen-Léopold, and call for the creation of the "new European neighborhoods" ("les nouveaux quartiers de l'Europe") in a conveniently located but vacant urban space on the second beltway of the city (the boulevard Lambermont, which further south becomes the boulevard St. Michel) (fig. 15).[16] The plan emphasizes the cultural and aesthetic gains for the city and Europe produced by an EC administrative/office park that has well-integrated residential and commercial uses in its morphology and draws its iconography from the rich stock of European architecture. Anticipating postmodernism, ARAU and IEB claim that such a plan would clearly demonstrate that the (classical) urban tradition of neighborhoods, roads, squares, and

---

15. Jean-Pierre Thiry, ed., *L'Europe à Bruxelles* (Brussels: Centre d'études et de recherche urbaines—ERU a.s.b.l.), p. 163.

16. ARAU/Commission Française de la Culture, "Un projet culturel pour l'Europe," in *Bruxelles vu par ses habitants* (Brussels: ARAU, 1984), pp. 124–27.

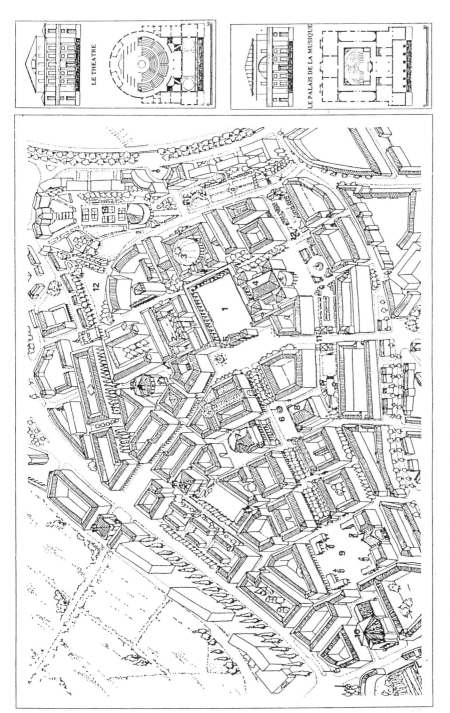

LE THEATRE

LE PALAIS DE LA MUSIQUE

Fig. 15. Plan of an alternative EC administrative park at the gare Josaphat

monuments in the style of the place Stanislas in Nancy or the place des Vosges in Paris would fully serve the needs of the European Communities for efficiency and security, and would be representative of European culture.[17] Neither the EC nor the Belgian authorities nor the private sector has taken these plans seriously, but they at least merit some consideration for their insistence on timeless imagery for a city and a quartier which may very well become the political heart of Europe.[18]

Intense place competition for the institutions of the European Communities and accompanying political-economic industries has made serious consideration of the four possible courses of action imperative for the Belgian political authorities and for the economic actors in the quartier Européen-Léopold. Market forces, economic expediency, and private initiative (essentially aspects of the same thing) appear to control the evolution of the quartiers Léopold and Nord-Est into the quartier Européen-Léopold. Despite lip service paid by the government to the preservation of multifunctionalism in the quartier, the development of new space, the redevelopment/gentrification of existing space, and the resulting land uses reproduce building types prevalent in American central business districts. The reproduced land uses differ from the American models, as does the raison d'être for the quartier; this is what makes the quartier Européen-Léopold a unique CBD—a CED.

---

17. Ibid., p. 124.

18. The Josaphat Station plan was still alive in 1991. ARAU/IEB restated their position on an eccentric EC administrative/office park and criticized the styling and location of the new EC extensions in the quartier Européen-Léopold in a press release. Hervé Cnudde, "Le projet ARAU de construction d'un second pole Européen sur la gare Josaphat pris en compte par les pouvoir public," press release, Brussels, 19 September 1991.

5

The Quartier as Land Market

T HE ORIGINAL quartier Léopold, conceived by the great
builder of Brussels Leopold II as a serene, upper-crust residential neighborhood, bears little resemblance to the quartier
Léopold of 1957–92, which we have renamed quartier Européen-
Léopold. Of the neighborhood's majestic mansions and palatial public buildings, there remain only traces: the southwest and southeast
sides of the square Frère Orban, and a few isolated mansions and
clusters of townhouses along the rue de la Loi, the rue Belliard, the
rue Joseph II, and the rues Guimard, Toulouse, and Pascale (fig. 16).
Less radical changes of a similar nature have taken place in areas
contiguous to the southeast (quartier St. Pierre, south of the parc du
Cinquantenaire), and to the north-northeast (quartier Nord-Est). Since
1957, market forces have been transforming in a piecemeal way the
residential landscape of mansions and eclectic, compact, fin-de siècle
townhouses into a morphologically and functionally modern and
postmodern central business district, by injecting in it high-value
administrative and office uses. This transformation translates into
billions of Belgian francs of business available to the city's economy:
BF 11.2 billion or U.S.$3.5 billion per year, every year.[1] Indeed a
perpetual Olympiad!

In this chapter, I will investigate the impact of the land market on
the physical and functional form of the quartier Européen-Léopold. I
will first review the fundamental concepts that drive land markets in
free market economies in general and the Brussels land market in
particular. Second, I will consider the different kinds of market parti-

---

1. Th. de Meulenaer, "L'Europe, ça rapporte . . . ," *Vlan* 1361 (19 December 1990): 16.
Very precise figures concerning the fixed expenses of the European Communities in
Brussels (rentals of buildings and leasing of business infrastructure, utilities, and salaries)
have been compiled by the think tank Mens en Ruimte: *Brussel, de internationale uitdag-
ing: De direct sociaal-economische impact van de internationale basisinstellingen in Brussels*
(Brussels: Mens en Ruimte v.z.w., 1990), pp. 5–29.

Fig. 16. The square Frère Orban. *Top:* The Church of St. Joseph flanked by restored neo-classical mansions (south side). *Bottom:* Newly constructed office buildings occupy the north side.

cipants by discussing the strengths and weaknesses of Logan and Molotch's framework on agency and market competition in Western cities. Third, I will describe and discuss the land market of the quartier by considering its relationship to other markets abroad and to submarkets in Brussels, and its land-use morphology. Fourth, I will present and discuss the findings of my opinion survey of real-estate development firms and independent real-estate agents on the land value profile, the direction of physical expansion of the business/administrative sector of the quartier, and the structure of market demand.

The Tools for Understanding a Capitalist Land Market

Land markets regulate the use of space in both urban and rural environments. Here we are concerned exclusively with the urban environment. Market forces forged by the supply and demand for vacant or improved land serve to allocate land among different uses and to establish gradients of land values in cities.

Land markets are intended to allocate the choicest pieces of real estate to the highest bidder. The highest bidder willingly pays the high price for locating his or her activities on a given central site because of the special advantages of that centrality. The income generated by this transaction is of great importance to local, Regional, and national plans for economic development, since cities and systems of cities contribute significantly to the accumulation of wealth and income. If we accept that land markets perform the land allocation task with perfect efficiency, turning out the highest income possible in every site, then it is in the interest of governments, which are one of the beneficiaries of the process through taxation, to interfere as little as possible with the operation of these markets. This normative notion of an intervention-free land market assumes that all the actors involved are economic actors, and are driven by profit maximization, and that the built environment in which they interact is easily adaptable to the tension between supply and demand.

Two principles guide urban land prices and urban land use: First, for any parcel, land price is what the highest bidder is willing to pay. Second, the more expensive the land, the greater are the incentives to economize on its use through more intense, or higher-density, usage.[2]

---

2. Mills and Hamilton, pp. 68–77.

There are clearly advantages to locating business offices close to those with whom there is interaction. The desire for accessibility to goods and services and the need to minimize transportation costs have given rise to central districts, and increasingly to additional eccentric nodes in Western cities. Assuming that everyone prefers a location close to the central district, bids for land rise; and since land prices go up as a result, land users and developers become willing to exchange greater urban densities for shorter distances from the central district.[3]

A look at the structure of U.S. downtowns confirms that business uses generally outbid residential uses. The result is the forming of a central node (CBD), a dense aggregation of similar high-rent business uses and the expelling of lower-rent uses, such as housing and large land consumers (e.g., hospitals, stadiums, cemeteries), to areas beyond the CBD, all the way to the urban fringe.

European cities are more complex and less adaptable to supply-and-demand dynamics because of ancient urban cores which may have emerged from nonmarket forces, the different valuation of traditional urban space and old buildings, and the existence of regulatory regimes which are not usually driven by profit maximization. Of course, these provisos bear a different significance in each European city. We will see that Brussels is atypical in this respect.

Density in both American and European cities is frequently historically determined. As George Tolley and Shou-yi Hao note, "The density near the center of the city is largely determined by the transportation costs that prevailed when that part of the city was first built up. If there is remaining vacant land, or if economic pressures are sufficiently strong to warrant demolition of old buildings, there will be some change in density in parts of the city already built up. Still, given that not all buildings are torn down, historical factors will influence density."[4] Indeed, downtowns of older cities are often a patchwork of older and newer buildings, some tall and some low, lining both rectilinear avenues, roundabouts with avenues in starburst formation, and narrower, sinuous streets. Controlling for all structural constraints to market forces, such as special zoning regulations, in theory, the land market determines when it is worthwhile to keep old buildings, or bear the cost of demolishing them and replac-

---

3. According to the model, if the price of land goes up, less will be used.

4. George S. Tolley and Shou-yi Hao, "Urban Land Use and Land Prices in Market Economies," typsescript, 1992, p. 13.

ing them with new and presumably denser (taller) buildings.

The magnet for density in cities in market economies is the existence of amenities: Tolley and Hao classify amenities as follows:

> A first type . . . is access or travel savings value and primarily reflects advantages of locating activities near to each other to save on travel and transport costs.
>
> A second type . . . consists of intrinsic characteristics of a location. The characteristics may be an attribute of the physical environment, a view or lack of air pollution, or it may be a cultural attribute as for example the characteristics of the people who live at or near the location.
>
> A third type of amenity is produced or affected by deliberate human effort, often by government. Examples are services from streets and other infrastructures, as well as schools.[5]

Discussing the impact of amenities on the location and growth of the Chicago office sector, Kenneth Posner considered three types of influences on 1991 CBD office rents: first, distance from the peak rent location at LaSalle and Madison Streets; second, building characteristics, such as size as measured by total rentable area, height of building, building age, years since last renovation, presence or absence of security personnel, architectural distinction, and other amenities such as shopping, health clubs, banking, and restaurants; third, locational effects such as proximity to prestigious shopping ("Magnificent Mile"), landmark buildings with Lake Michigan and Loop views, and other significant parts of the Loop. Posner also confirmed that certain prestigious buildings, such as the Sears Tower and John Hancock Tower, act as subcenters in the CBD.[6] This special pattern of urban development around prestigious "anchor buildings" is of special significance in the quartier Européen-Léopold, where next to the public amenities listed by Tolley, Hao, and Posner, the landmark buildings Charlemagne, Berlaymont, the new Council of Ministers facility, and the Centre International du Congrés act as anchors for urban development.

---

5. Ibid., box 4: "Amenities."

6. Kenneth Posner, "An Amenity Analysis of Office Rents in the Chicago Central Business District," *Journal of the American Real Estate and Urban Economics Association* (forthcoming).

Who Represents the Interests of the Private Sector?

Logan and Molotch in their study of the interaction of exchange and use values argue that cities are interest-driven social constructs. Land markets work as social phenomena, and fundamental to the understanding of their function "are the social contexts through which [land and buildings] are used and exchanged." They note that every piece of property has both a use value (for example, for the person who calls it "home"), and an exchange value for having the capacity to produce revenue.[7] For Logan and Molotch it is the stresses and conflict between exchange and use values of properties that shape the cities, as different individuals and groups of agents embrace one over the other. The end result is a process by which city wealth is created and harvested by complexes of elites. This framework of agency in an urban land market offers some basis for understanding who is participating in controlling the Brussels land market.

The cortex of Logan and Molotch's model land market is made up of growth coalitions, which are loosely allied, self-interested, profit-maximizing "place entrepreneurs": "people directly involved in the exchange of places and the collection of rents, [who] have the job of trapping human activity at the sites of their pecuniary interests."[8] Place entrepreneurs are both catalysts and instruments of the market. They provide the rationale for movements at all levels as well as the means to carry them out. They are both the hunter and the victim of the real-estate game. They occupy the vast territory between institutional players of the private and the public sectors.

Logan and Molotch distinguish three kinds of place entrepreneurs, each with a different social relationship to place and each having a distinct influence and role in the system of use and exchange:[9]

- Serendipitous entrepreneurs: These are the typical *rentiers* who harvest the revenues from one or more real-estate properties without having played an active role in acquiring, developing, or using the property to launch or capitalize other land projects. Lacking in motivation and sophistication, serendipitous entrepreneurs are passive participants and can become the prey of place entrepreneurs of an opportunistic character.

---

7. John R. Logan and Harvey L. Molotch, *Urban Fortunes: The Political Economy of Place* (Berkeley: University of California Press, 1987), pp. 1–2.

8. Ibid., p. 29.

9. Ibid.

- Active entrepreneurs: These are fairly likely to exploit the serendipitous entrepreneurs. They actively research and purchase properties in locations likely to appreciate. They can be likened to antennae tuning in on developments in local politics and economics in order to discern the best marketing strategy. While they use their investigative skills in estimating the geographical movements of others, their resources are limited by the fact that they have access only to local social networks.
- Structural speculators: These are the most important among place entrepreneurs. They are usually the business elite. They not only strategize on the basis of market information but help structure the markets in ways that benefit them. They possess political and often economic clout and influence planning regulations, political appointments, and locational decisions concerning key city projects. Unlike active entrepreneurs who depend on acute and assiduous observation of the market, structural speculators possess the equivalent of insider information.[10]

This classification of agency suggests a gradient of influence on markets which assumes that some players are ultimately subordinate to other players. The commodity for which the players compete is as much information as it is land and buildings.

Of the three kinds of place entrepreneurs, the ones who matter in the constitution of a "growth coalition," which promotes the growth of a city in the direction of its own interests, are the structural speculators. In ideal terms, they cooperate with politicians, the local media, and leaders of "independent" public or quasi-public agencies, such as utilities, to formulate a vision and strategy for the city they are promoting.[11] Implicit in this system is the understanding that each party is at the same time promoting its own individual interest, resulting in the promotion of certain sites and locations in the city over others. Serendipitous and active entrepreneurs are consequently required to make adjustments to these movements. Some will be successful in adapting while others may choose or be forced to leave the market.

Against these entrepreneurs is arrayed a valiant ragtag assemblage of persons who embrace use-value over profit maximization, and who coalesce to protect their neighborly way of life and liveli-

---

10. Ibid., pp. 29–31.
11. Ibid., pp. 62–74.

hood against the excesses and adventurism of the exchange-value machine.

Logan and Molotch's presentation of how "urban fortunes" of certain place-based elites are constructed is compelling and offers enticing "real world" detail. Yet, their agency-driven model of accumulation fails considerably in addressing certain important analytical imperatives.

First, without making clear the position of their market model within neoclassical land-rent theory, Logan and Molotch make explicit neoclassical economic assumptions about the processes that drive growth and influence prices. With these assumptions they saddle their framework with the normative baggage of the neoclassical market model. While Gordon Clark praises the study for its cultural realism and its exploration of the sobering aspects of the capitalist marketing of urban land,[12] how, he asks, can this clear strength fail to be compromised by the "unreality" of the underlying framework? Logan defends the framework by noting that despite the neoclassical market assumptions it makes, his model differs dramatically from conventional neoclassical economics and human ecology by focusing both on people as residents, and on different consumers of land as people.[13]

Second, the agents promoting or resisting exchange-value accumulation appear monolithic. In a free market society, many members of the population find themselves landlord and renter at different time periods, or even at the same time. This duality of interest resident in most persons is not explored, despite the implicit behavioral analytical aspirations the study entertains. Robert Lake is distinctly non-Marxist in arguing that the use and exchange value distinction comprises a "permeable duality" rather than a sharp dichotomy.[14] Moreover, the profiles of the agents have been criticized as caricatures: Lake argues that the lack of context qua structural analysis leads Logan and Molotch to draw their characters out of proportion: "The local rentiers that fuel the growth machine have astonishing reach,"[15] a claim that remains unsupported by empirical evidence.

---

12. Gordon L. Clark, "A Realist Project: *Urban Fortunes: The Political Economy of Place*," *Urban Geography* 11, no. 2 (1990): 196.

13. John Logan, "How to Study the City: Arguments for a New Approach," *Urban Geography* 11, no. 2 (1990): 201.

14. Robert Lake, "*Urban Fortunes: The Political Economy of Place:* A Commentary," *Urban Geography* 11, no. 2 (1990): 182.

15. Lake, p. 180.

David Harvey also finds fault in the stratified classification of agents for failing to discuss the role of class and faction formation.[16]

Third, Logan and Molotch's very sharp focus on agency has drawn criticism from scholars who believe that structural causes are critical to explaining exchange-value-driven urban fortunes and the power of urban growth machines. "Agents beg for a context," Robert Lake claims. "Indeed, particularly after Giddens, the concept of agency has no meaning divorced from structure. The case for an 'interest-driven social construction of cities' will always be incomplete if it ignores the prior questions of how and why interests are established."[17] Lake refers, of course, to a capitalist structure and the reproduction of capitalist relations. Susan Clarke adds to this criticism, noting that *Urban Fortunes* pays little attention to important shifting structures such as the new international division of labor, qualitative changes in the mode and organization of production, the growing importance of information technologies, and the expansion of informal economies in developed and developing countries,[18] except in general terms: "the continuing evolution of the placeness of wealth."[19] Clearly, these evolving features of corporate organization should have an impact on the structure and character of the urban growth coalitions. Hence, "the political economy of place" cannot be generalized in the terms in which Logan and Molotch present it.

What importance does Logan and Molotch's model have for our discussion of the construction of European space in the city of Brussels? While the present investigation of Brussels urbanism differs in its method and findings, Logan and Molotch's framework constitutes a remarkably appropriate focusing lens for some of the same general preoccupations that drive this study. The challenge we face here is that of avoiding the shortcomings of Logan and Molotch in formulating an alternative political, economic, and aesthetic framework for understanding the role of European place/space construction in Brussels urbanization. This task will be addressed in this and the following two chapters.

In general terms, several points of connection exist between the

---

16. David Harvey, "*Urban Fortunes: The Political Economy of Place:* Review," *Environment and Planning D: Society and Space* 8 (1990): 495–96.

17. Lake, pp. 180–82. In his commentary, he refers to Anthony Giddens's *The Constitution of Society* (Berkeley: University of California Press, 1984).

18. Susan E. Clarke, "'Precious Place: The Local Growth Machine in an Era of Global Restructuring," *Urban Geography* 11, no. 2 (1990): 186–87.

19. Logan and Molotch, p. 253.

framework presented in *Urban Fortunes* and the present framework applied to Brussels: Brussels as a land market functions in terms of capitalist principles. In the next section, I explore the Brussels land market in terms of the principles introduced in the first section of this chapter. Its structure and trends are further qualified and explained in more than economic terms in the following chapters.

The quartier Européen-Léopold, as an area in flux between a quaint old residential neighborhood and world-class real estate, hosts the full spectrum of place entrepreneurs: some want to preserve the quaint residential Brussels character, while others want to profit from the fortuitous presence of the EC institutions in the quartier. The tensions and struggle between the local citizens' committees and the growth coalition are receiving an appreciable amount of press. One can clearly identify a contest of competing interests in the quartier.

The city, Regional, and federal governments have a stake in keeping the European Communities in Brussels and making a success of the quartier Européen-Léopold as an international administrative and business CBD. There is firm evidence that a growth coalition guiding urban development has emerged.

Logan and Molotch underline the significance of state and local government in mitigating the process of accumulation. In the case of the quartier Européen-Léopold, as in the generic city presented in *Urban Fortunes*, there appears to exist tension between the democratic process, government policy, and the aspirations of members of the growth coalition.

Most important, *Urban Fortunes* raises questions about agency in urban process, and contributes, albeit deficiently, to the debate over the appropriate formula for combining agency and structure in interpretive social science modeling. *Urban Fortunes* leans heavily on the side of agency. Logan notes that it "emphasizes action, purpose, and the visible consequences of decisions; it seeks to impute no consequences to structures without identifying the agents through which those effects come about."[20]

This study attempts to balance the two, not out of timidity but because of firm evidence indicating that structures, such as the process of managed complex interstate convergence as expressed by the European Union, have a determining role to play in the concrete ambitions of the players of the quartier.

---

20. Logan, p. 202.

Brussels in Competition with Other Urban Centers

Brussels still represents a lacuna in office space rental values. It has yet to command the number of headquarters of world-class corporations that London and Paris command, though its position is improving. Its stock market, though evolving toward liberalization, is less dynamic, less affluent, less technologically up to date, and less well linked to other markets than those of London, Paris, and Frankfurt, to name only the European competition. According to 1989 data, in a ranking of the top twenty-eight cities for business on the basis of total rental costs per square meter, Brussels ranked third from the bottom (table 1).

Perhaps competitively priced office space is the reason that Brussels became one of the hottest real-estate markets of the late 1980s. It has relied on this price competitiveness, on very solid fiscal incentives for corporations seeking a Regional or global headquarters, and on its appeal as a smaller, more livable city to attract corporate clients. Morgan Guarantee is currently preparing to move its worldwide clearinghouse operations to its new skyscraper in the quartier Nord. IBM has its largest Continental facility at La Hulpe, an attractive suburban community a few miles outside Brussels. Other corporations establish large operations along the boulevard de la Woluwe, while reserving the front-office operations for the quartier Européen-Léopold and other prestigious addresses in the city. Adding the capital surpluses of the late 1980s to the successful launching of the Single European Market program has helped Brussels real estate reach these relatively modest levels, which are high by comparison to the past (fig. 17).

The most recent data have been compiled by the international realty firm Healey & Baker in the third edition of its *European Real Estate Monitor*.[21] The realty firm commissioned the Harris Research Center of London to conduct a poll of firms on their preferences for and attitudes toward the main business centers of Europe. The sample of 530 business executives from nine countries was drawn from a population of fifteen thousand of the largest European corporations. The findings were generally very positive for Brussels, and confirmed it as a world-class business center: it rated forth in Europe be

---

21. Healey & Baker, *European Real Estate Monitor*, 3d ed. (Brussels: Healey & Baker, 1992); reported in detail in Owen Bradley, "Etre ou ne pas être à Bruxelles?" *L'évenement immobilièr* 72 (December 1992): 7.

TABLE 1. Rental prices of prime office space in December 1989, in U.S.$ per m^2

Tokyo	2,191	Barcelona	476
London, City	1,640	Beijing	474
London, West End	1,473	Chicago	474
Hong-Kong	1,131	Lisbon	444
Paris	860	Manchester	438
New York, Midtown	690	Glasgow	437
Milan	633	Washington, DC	430
Madrid	625	Los Angeles	416
Singapore	574	Melbourne	408
Sydney	574	Perth	380
Stockholm	560	Sao Paolo	308
Frankfurt	513	**Brussels**	**280**
New York, Downtown	509	Bangkok	221
Toronto	479	Amsterdam	220

Sources: Richard Ellis [real-estate development corporation], *Niveaux des loyers dans le monde* (Brussels: Richard Ellis, 1989). Jones Lang Wootton International carried out a similar survey confirming the low ranking of Brussels office costs. Jones Lang Wootton International, "Worldwide Office Occupancy Costs," *JLW Research* 1 (September 1989): 1.

hind the traditional world-class business centers of London, Paris, and Frankfurt. It ranked ahead of Amsterdam (5th), Geneva (10th), Berlin (14th), Munich (16th), Madrid (also 16th) and Moscow (26th).

The criteria used by the polling firm were closely related to the general needs of firms researching new business sites and not necessarily to the "Europeanness" of the city, that is, its closeness to EC institutions. At the same time, however, firms must have taken Brussels's special relation to the Communities into consideration in weighing their answers. Specifically, the criteria ranked below in order of importance are deemed essential to the macrolocation decisions of firms. The number of respondents giving Brussels the highest score in each of these areas was as follows:

- Access to markets (62%): Brussels ranked 4th overall
- Quality of international transportation links (49%)
- Quality of telecommunications (43%): ranked 4th overall[22]
- The cost and availability of labor (39%): ranked 13th overall
- The political climate, including fiscal incentives (34%): ranked 2d overall
- The profitability of office space (23%): ranked 5th overall, be-

---

22. This is not a tribute to the good quality of Brussels's telecommunications, but an indication of how inefficient state-run telecommunications are Europewide.

hind Glasgow, Lisbon, Manchester, and Prague
- The availability of office space (22%): ranked 6th overall, behind London, Paris, Manchester, Frankfurt, and Glasgow
- The ease of internal mobility (22%): ranked 5th overall
- The number of spoken languages (17%): ranked ahead of London and Amsterdam
- The absence of pollution (11%): ranked 10th overall
- The quality of life of the executive (10%): ranked 15th overall

Brussels scored very well in areas of importance to firms, such as access to markets and quality of infrastructure, and poorly in areas of less interest to firms, such as quality of life. This should be of great interest to observers of urban affairs in Brussels. Factors such as quality of life and pollution may be of lesser significance to business location decision making but of great significance for full-time Brussels residents. Firms constitute an important lobby in Brussels—as in other business centers—and can have critical input in land-use decisions made by the Brussels Regional government. The fact that quality of life is rated low suggests that the corporate lobby of Brussels is little inclined to promote significant changes that many feel are needed on that front.[23]

Fifty percent of the firms polled already have a presence in Brussels, as compared with 64 percent and 62 percent present in Paris and London respectively. An additional 6 percent of respondents whose firms do not have a presence in Brussels currently, plan to open offices there in the next five years.

Fifty-nine percent of the respondents—up from 52 percent in 1990 and 55 percent in 1991—estimate that Brussels will become the city of principal political significance within the next five years, well ahead of Berlin (1990: 31 percent; 1991: 25 percent; 1992: 23 percent) and Paris (1990: 10 percent; 1991: 13 percent; 1992: 11 percent). Strasbourg, which is Brussels's main rival for the European Parliament, attracted only 4 percent positive responses to the related request to "name other cities which could have an impact in the next five years on the business world."

Strasbourg's claim to the European Parliament is supported by an early decision to hold EP plenary sessions there. Strasbourg constructed a modern suburban facility, including a large amphitheater

---

23. Later in this chapter, I will present the findings of a survey investigating the views of urban development professionals on the quartier Européen-Léopold. Some of the findings of the Healey & Baker macrostudy are echoed by this microstudy of the quartier.

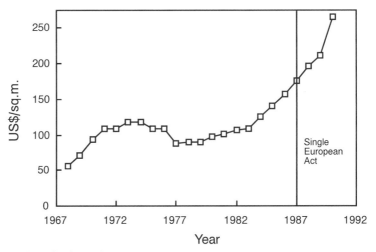

Fig. 17. Growth of rental properties at the quartier Européen-Leopold. Adapted from Healey & Baker [real-estate development corporation], *Belgium* (Brussels: Healey & Baker, 1990), p. 4.

for that purpose. It is now constructing a new, larger facility and amphitheater in competition with the Centre International du Congrés of Brussels. Strasbourg Mayor Catherine Trautmann and French President François Mitterand, agreeing that the Parliament belongs to Strasbourg and should remain there at all costs, launched a public relations campaign in January 1989. In an effort to strong-arm its European partners into conceding the European Parliament to Strasbourg, France is refusing to participate in the voting to designate where other Community institutions will be housed.

Strasbourg's claim is also supported by other intangibles: as the capital of Alsace, it belongs to and is a product of interaction and competition between the French and German cultures. If we accept the somewhat simplified view of European integration as the product of Franco-German aspirations for Continental peace and prosperity, then Alsace and Strasbourg become appropriate symbolic places for the nurturing of a pan-European democratic regime.

As expected, the Belgian government actively promotes Brussels's claim to the European Parliament. The key argument is one of efficiency: the annual bill for the shuttling of persons and documents between Brussels, Luxembourg, and Strasbourg is between 1.3 and 1.7 billion Belgian francs (ca. U.S.$42 million, or 10 percent of the Parliament's budget). The distribution of EP activities between these three cities requires the regular movement of approximately 3,200 persons and 1,000 trunks of documents.[24] In addition to avoiding these logistical complications, permanent establishment of the EP in Strasbourg would result in significant financial savings, as the rental cost of the Strasbourg facilities is lower by approximately 18 percent (100 vs. 115 ecu per m^2).[25] Not to have chosen Strasbourg has rightly been called absurd. The poor rating of Strasbourg in the Healey & Baker survey indicates that the business world does not share the preoccupations of the French political leadership and appears increasingly attached to Brussels.

Quartier Européen-Léopold vs. Other Brussels Submarkets

The real-estate industry distinguishes four major kinds of office space consumers, and a number of central nodes of tertiary activity (table 2).

TABLE 2. Main office space consumers, in square miles

Belgian state	2,100,000
International organizations (EEC)	1,000,000
Belgian private sector	1,300,000
Foreign private sector	1,600,000
Vacant	200,000

Source: Healey & Baker, *Belgium* (Brussels: Healey & Baker, 1990), p. 10.

Unlike cities like Chicago which have a clearly definable central business district and a number of recent suburban business districts which concentrate the bulk of administrative/office space, Brussels has a number of office districts in the city and lately in its periphery. In simple terms, its CBD is made up of a number of fragments. Until the early 1960s, the Pentagon city concentrated the majority of administrative/office uses. Since that time, the area we define as the quartier Européen-Léopold has become the single most compact district for administrative/office uses. Local real-estate developers have come to call it "the CBD" because it comes closest to an American-style CBD in terms of density of use and building types. The fast-growing quartier Nord, conceived in the late 1960s as the

---

24. Paul Staes, *Europe at Stake in the Speculation Game: A Political and Ecological Analysis of a Billion-Franc Project; The European Parliament Plays Pontius Pilate* (Brussels: European Parliament document no. 95374, September 1990), p. 32.

25. *European Parliament Building Service Statistics on Space Availability and Cost in Luxembourg, Strasbourg, and Brussels* (Brussels: EP document no. 131.758/4).

"Manhattan Project" by real-estate mogul Charlie De Pauw, features a cluster of the tallest office towers in Brussels and may soon challenge the quartier Européen-Léopold for the title of American-style CBD.[26] Other concentrations of office space can be found along the avenue Louise, and increasingly in the periphery north-northeast, east, and southeast of the city (fig. 18).

The Brussels administrative/office sector has been growing rapidly since the 1960s. According to a study of the executive of the Brussels-Capital Region, "the total area occupied by the office sector tripled between 1966 and 1988, and doubled between 1973 and 1988.[27] As of January 1990, Brussels with 6,200,000 m^2 represented 75 percent of total office space in Belgium. By 1992, Brussels office space rose to more than 7,000,000 m^2" (table 3).

The new buildings leased by the EC executive and the legislative and other collateral agencies will add at least an another 500,000 m^2 to the 1,930,800 m^2 of the quartier Léopold and Schuman by the end of 1995.[28] The data in table 3 illustrate two trends: first, the intensification of real-estate investment in administrative/office space in the quartier Européen-Léopold since 1985, and second the decentralization of administrative/office uses to the suburbs. In the former case, seven years of real-estate development have added at least 430,000 m^2 of administrative/office space to the quartier Européen-Léopold, which in 1992 already concentrated 27.5 percent of the total administrative/office space in the Brussels metropolitan area. The latter case is exemplified by the relative price performance of the various areas of the city and its periphery which attract the office trade (table 4).

While the quartier Européen-Léopold led all urban areas in price growth (+6.22 percent), the suburban zones directly east and southeast of the center (boulevard de la Woluwe, the avenue de Tervueren, and the boulevard du Souverain) surpass its performance by 50 percent.

---

26. It is important to note here that the resemblance of these Brussels office-dominated quartiers to the typical American CBD end with the consideration of aesthetic.

27. A. M. Vandenbossche, *Les bureaux à Bruxelles* (Brussels: Ministère de la Région Bruxelloise, bureau d'architecture, 1988).

28. In table 3, the Regional government distinguishes two subareas of the quartier Européen-Léopold: quartiers Léopold and Schuman. A more modest projection by the Brussels-Capital Region predicts that in 1989–2008 an average of 100,000 m^2 will be added annually to the city's total stock. Clearly, it does not take into account the megaprojects which were given the green light by the government shortly after that study was published. Ibid.

Fig. 18. The Manhattan-style quartier Nord. The boulevard du Jardin Botanique separates the Pentagon from the quartier Nord. The Sheraton Hotel and the recently completed "euroclearing" facility of Morgan Guarantee appear in the background.

TABLE 3. Brussels office submarkets

	As of 31/12/84		As of 1/7/92	
	No. of bldgs.	Surface (m²)	No. of bldgs.	Surface (m²)
Pentagon	217	1,714,200	235	1,887,700
Q. Léopold	214	1,125,900	250	1,386,800
Schuman	34	374,500	42	544,000
avenue Louise	67	390,300	73	446,000
Laeken	5	9,000	6	13,000
Haren	3	23,750	10	68,500
Q. Nord	10	176,000	11	232,000
Totals, Brussels municipality	550	3,813,650	627	4,578,000
St.-Josse	29	277,300	33	399,400
Schaerbeek	32	128,400	34	142,600
Etterbeek	16	131,700	20	167,700
Ixelles	50	239,300	55	263,000
St.-Gilles	42	222,300	47	258,000
Totals, 1st Beltway	169	999,000	189	1,230,700
Evere	13	92,200	26	178,100
Woluwe-St. Lambert	13	62,100	35	267,300
Woluwe-St. Pierre	13	110,800	16	125,000
Auderghem	12	47,300	40	215,900
Watermael-Boits.	17	171,200	19	199,200
Uccle	22	88,400	23	91,400
Totals, 2nd Beltway	90	572,000	159	1,076,900
West Brussels	28	128,500	31	172,600
Grand Totals	837	5,513,150	1006	7,058,200

Source: Region de Bruxelles-Capitale, *Stratégie pour une géstion regionale de la fonction administrative*, Administration de l'urbanisme et de l'aménagement du territoire, Document de synthèse (Brussels: BRAT, 1991,) p. 2.

The important question concerning the maintenance of these trends is whether office space buyers and lessors can readily substitute one location for the other. It is also unclear how differently the key amenities of each location will be evaluated by market players: on one hand, the quartier Européen-Léopold is uniquely situated vis-à-vis the European Communities, while on the other hand the sub-

	In new buildings		In surface area built		
			Growth between 1/1/85 and 1/7/92		
No.	% growth in no.	% growth in grand total	New, built (m²)	% growth	% growth in grand total of new surface built[a]
18	8.2	11	173,500	10.1	11
36	16.8	21	260,900	23.1	17
8	23.5	5	169,500	45.2	11
6	8.9	4	55,700	14.2	4
1	20.0	1	4,000	44.4	0
7	233.3	4	44,750	188.4	3
1	10.0	1	56,000	31.8	4
77	14.0	46	764,350	20.0	20
4	13.7	2	122,100	44.0	8
2	6.2	1	14,200	11.0	1
4	25.0	2	36,000	27.3	2
5	10.0	3	23,700	9.9	2
5	11.9	3	35,700	16.0	2
20	11.8	12	231,700	23.1	15
13	100.0	8	85,900	93.1	6
22	169.2	13	205,200	330.4	13
3	23.0	2	14,200	12.8	1
28	233.3	17	168,600	356.4	11
2	11.7	1	28,000	16.3	2
1	4.5	1	3,000	3.3	0
69	76.6	41	504,900	88.2	33
3	10.7	2	44,100	34.3	3
169	20.1	100	1,545,050	28.0	100

[a]1,545.050m²

urban node provides modern, first-rate office construction, as well as release from urban externalities such as pollution and traffic congestion, and is relatively unencumbered by the ever more restrictive regulations of the city of Brussels. Judging by rental prices alone, for the moment the quartier Européen-Léopold is in greater demand in spite of yielding the lowest rate of return in comparison with the other business nodes of the city (table 5).

Increasingly, however, the attention of urban developers has

TABLE 4. Rental price increases by business node for 1991 (in %)

Pentagon	+ 1.45
Quartier Européen-Léopold	+ 6.22
Quartier Nord	+ 2.88
Avenue Louise	+ 1.00
Suburban (between the *petite ceinture* and the beltway: e.g., boulevard de la Woluwe)	+ 9.34
Beltway	− 2.83

Source: Office des Propriétaires Immobilier S.A., *Le Marché Bruxellois* (Brussels: Etude de la société OP, 1992), p. 10.

TABLE 5. Range of rental prices and percentage rates of return for office space $(BF/m^2/yr)$

	Yield (%)		Rent (BF)	
Business nodes	From	To	From	To
Pentagon	7.40	7.75	5,800	7,000
Quartier Nord	7.25	7.50	7,800	9,500
Quartier Européen-Léopold	**7.00**	**7.25**	**6,800**	**10,000**
Avenue Louise	8.00	8.25	5,500	8,000
Suburban	**7.75**	**8.25**	**6,000**	**8,000**
Beltway	8.75	9.50	4,500	6,500

Source: Office des Propriétaires Immobilier S.A., *Le Marché Bruxellois* (Brussels: Etude de la société OP, 1992), p. 11.

TABLE 6. Demand for office space by business node (in m^2)

Business nodes	July 1991	July 1992
Pentagon	31,000	87,000
Quartier Nord	114,000	21,750
Quartier Européen-Léopold	100,000	68,500
Avenue/quartier Louise	17,000	11,000
Suburban	190,000	166,750
Beltway	28,000	39,000
TOTAL	480,000	394,000

Source: M. J. Bamber, ed., "Property Report Belgium," *Richard Ellis Research* (September 1992): 2.

turned to the suburban business area. Despite the 1990 recession's dampening effect on new construction and demand for office space, the suburbs appear to lead the quartier Européen-Léopold in demand (table 6).

Assuming the continuing presence of the European Communities in Brussels, three factors may have a profound impact on the substitution effect involving the quartier Européen-Léopold and the suburban business node:

- Improved surface or subway/rail communications between the suburban locations in question and the EC administrative park
- Exhaustion of available space for office building development inside the quartier Européen-Léopold such as:
  —Vacant building lots
  —Transformable older building stock (office or other)
  —Land dedicated to other uses
- Severe and enforced urban planning regulations:
  —Limiting the absorption of residential space
  —Limiting the density of building (e.g., the height and volume of single buildings)
  —Imposing an onerous fiscal regime on the office development industry, and on small developers illegally transforming residential space into office space

Any of these factors will increase the likelihood that office space consumers will substitute green suburban locations for the more expensive and congested quartier Européen-Léopold. Yet, as likely as these factors may appear, the short-term reality is that the quartier Européen-Léopold benefits from the comparative advantage of hosting the European Communities.

Mapping the Quartier: The Evolution of the Quartier's Real-estate Market

Whereas general information about the numerous real-estate markets within the metropolitan area is published regularly by banks and the large real-estate development firms for their investor clients, specific and reliable time series data on the pricing of properties are difficult to find. The data that are available to the public are not sufficient to provide statistically significant conclusions about the evolution of land values in the quartier. When the available data, however, are assessed in the light of corroborating industry reports on the markets of Brussels, important conclusions can indeed be extracted.

The database used here has been extracted from the 1980–91 annual reports of the association of Brussels notaries. A review of the

sale records of all publicly auctioned properties in the quartier
between 1980 and 1991 helps us distinguish, albeit in fairly impressionistic terms, the land value profile of and the trajectory of land
prices in the quartier. The annual reports list the address of each auctioned property, the building type and the number of stories (when a
building is involved), the surface area of the lot, the allowable land
use (as listed in the city's planning records), the property tax, the
price and date of sale, and finally, the name of the notary responsible
for bringing the property to market. The database includes 109 auctioned "lots" (115 quartier properties). Mean values for each of the
eleven years have been used to plot the evolution of the quartier's
real-estate market during a period of great socioeconomic change.[29]

The resulting picture is striking (fig. 19). The market appears to
have taken off during 1985 after idling for the first half of the decade.
After a five-year long period of extraordinary price increases, the
market fell sharply. The correlation of the market's take-off with
Jacques Delors's "relaunching" of Europe and the plans for a Single
European Market cannot be avoided. Likewise, the market's recession can be associated with the world economic tensions resulting
from the 1991 Gulf War, the onset of a general economic recession in
the United States that same year, and the Belgian "franc fort" monetary policy which kept interest rates high and capital expensive.[30]
Again, however, it is important to underline that the curve only
suggests the evolution, since the sample is not statistically significant
for all years—especially 1986–87 and 1989–91.[31]

A careful examination of the database yields the following basic

---

29. According to real-estate development professionals, properties brought to market
in an auction setting are usually sold for below-market prices. Although there are exceptions, the reader should estimate that regular sale prices would be 10–25 percent
higher than those listed in the notarial database.

30. The frequent withdrawl of properties from the market starting in 1990 was an
indication that real-estate speculators were unwilling to concede that the market was
overheated. Clearly, in the light of the deepening recession, the market could not sus-
tain the extraordinary growth rate in land values set between 1985 and 1990. Real-estate professionals suggested that by 1991 the quartier market was overpriced by as
much as 30 percent.

31. There are two sets of plotted points that determine each curve. The curve labeled "mean" is prescribed by points representing the mean score of the value/area ratios of all properties auctioned during each of the twelve years represented in the
database. The curve labeled "mean/central" is determined by points representing the
mean score of the value/area ratios of the *subset* of properties auctioned during a given
year which are situated in close proximity to the major axes of the quartier and major
EC facilities.

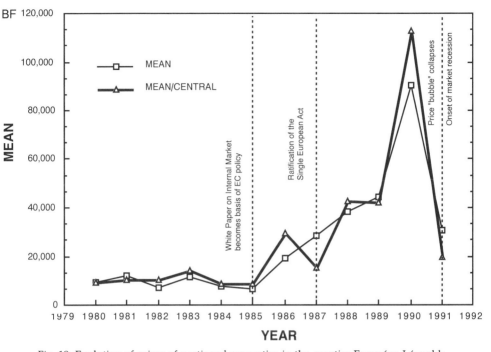

Fig. 19. Evolution of prices of auctioned properties in the quartier Européen-Léopold: 1980–1991

observations: more than 86 percent of the quartier properties auctioned during the 1980–91 period were rowhouses. Eighty percent of them constituted residential space and an additional 12 percent were shop houses, combining commercial space at street level with one or more apartments in the floors above. Most of these rowhouses were eventually demolished and replaced by office blocks or EC administration buildings. An exception constitutes the rowhouses on the rues de Pascale and Toulouse.

Fifty-five of the 115 properties auctioned were situated in a small number of streets: Belliard, des Confédérés, de Pascale, Joseph II, Livingston, Palmerston, Stévin, Toulouse, and Van Maerlant. All but the rue des Confédérés are situated within the core of the quartier.

According to the notarial records, the majority of the auctioned properties were not in use at the time of sale, which suggests that the level of vacancy and abandonment of nineteenth-century buildings in the quartier, especially until 1985, was very high.[32] Moreover, the

---

32. Whereas in the United States the owner of a building can demand the vacating

significant variance in value/area ratios within the same year, between similar styles of properties, on the same street, suggests that
nineteenth-century buildings vary widely in their present state of
repair. For example, the auction sale prices of four rowhouse properties in the eclectic style, dating to 1890, on rue Palmerston, varied by
as much as 33 percent (cases 14–17 in table 7).[33]

The rapid turnover of certain properties suggests that speculators
were operating in the market: for example, four of five residential
rowhouses on rue de Pascale, situated on the same city block as the
newly erected annex to the European Parliament,[34] were resold
within one to two years.[35] Three of the four transactions were profitable, yielding a return of 10–20 percent on the investment. Not all
transactions, however, are easy to explicate in market terms: the
rowhouse on rue Archimède 71 was auctioned on 26 October 1990 for
12,200,00 Belgian Francs. On 22 February 1991, it was again put on
the market at exactly 50 percent of the original sale price.[36]

The spatial distribution and the temporal spread of certain transactions suggest that buyers were attempting to consolidate small parcels into large lots: for example, between 13 January 1983 and 6
March 1984, the office of the notary Vernimmen auctioned in four different transactions nine contiguous properties forming the corner of
the rues Belliard and Van Maerlant. Although the identity of the
buyer or buyers is not revealed in the annual report, a visitor to that
street corner will discover today the newly erected annex of the European Parliament. A number of smaller sequences of properties in
highly valued locations can be discerned: 107–9 rue du Commerce,
17–23 rue Palmerston, 5–7 boulevard Livingston, and 111–19 rue
Stévin.[37]

Although the quartier's real-estate market may have appeared
overpriced to many by the end of 1990, one need only compare any
property in the quartier to the 139 million Belgian Franc (ca. U.S.

---

of the premises on very short notice, in Belgium laws vigorously protect renters' rights—
especially if the renter is an elderly person or a family. Therefore, the high rate of vacancy at the time of sale suggests that the buildings were mostly unoccupied.

[33] *Répertoire des ventes publiques réalisées pendant les année 1980–1991 par les notaires de
l'arrondissement de Bruxelles*, 11 vols. (Brussels: Editeur ventes notariales, 1980–1992),
1980, cases 14–17.

[34] Ibid., 1982: cases 29–33.

[35] Ibid., 1983: cases 41–42; 1984: cases 50–51.

[36] Ibid., 1990: case 96; 1991: case 103.

[37] Ibid., 1980: cases 6–7; 1980: cases 14–17; 1984: case 54; 1987: case 80; 1988: case 91,
respectively.

$560/m^2$) auction price of the mid-rise building on rue Belliard—a long-valued area close to the inner ring and the Royal Park.[38] The U.S.$560 per square meter price tag evokes the London City, Paris, or New York markets. By 1990, the rate of growth in the value of up-market properties in the quartier made Brussels one of the most sought-after real-estate markets in the world.

The consequences of these changes are clear: the property titles are leaving the hands of small owners (Logan and Molotch's "serendipitous entrepreneurs") and go to large-scale buyers, often corporate and institutional market players. The demographic and economic structures of the quartier are changing as residential properties give way to the office function. Finally, the morphological character of the quartier is changing rapidly with the removal of the nineteenth-century buildings. The auction records strongly corroborate these changes and shed more light on the gaming process (table 7).

Professional Opinions on the Structure of Demand

It is important to take stock of activity in the quartier as expressed by changing land use and market trends and perceptions. Data for this section were obtained through fieldwork in the quartier, from a mail-in survey of urban development firms and independent real-estate agents and from a series of interviews with experts and residents of the quartier. Time series property-level information on land values cannot be obtained for a statistically significant sample of properties in the city. Except for the aggregate figures presented above, information at the property level is confidential unless the owner/lessor/developer decides to disclose it for reasons of expediency. Price information can be obtained from classified advertisements for the sale of real-estate properties. This information is however generally misleading: international publicity around "1992" and the ensuing real-estate speculation caused the hyperinflation of prices during the real-estate boom of 1987–90. During this period, prized properties were often advertised at 30–100 percent over market value.[39] However, since this information is of great value in discerning the urban impact of European integration in this part of Brussels, I asked real-

---

[38] Ibid., 1990: case 97.

39. Interview with Mr. Harold Semal, general manager of Ascona & Co., Brussels, 3 March 1990.

TABLE 7. Auctioned properties in the Quartier Européen-Léopold, 1980–91

No.	Address	Type bldg. [a]	Land use	Area[b] (m^2)	Bf [c] (000)	Value[d] (m^2)
1980						
1	**35 Arion**	Rowhouse	Com/Res.	50	675	13,438[e]
2	24 Arts	Office block	Office	floor	800	?
3	64 Bordiau	Rowhouse	Residential	123	910	7,398
4	**33 Breydel**	Rowhouse	Residential	100	780	7,800
5	**49 C. Martel**	Rowhouse	Residential	55	450	8,181
6	**107 Commerce**	Rowh./MM	Residential	308		
	109 Commerce	Rowh./MM	Residential	288	13,000	21,812
7	**66 Wautier**	Rowhouse	Commercial	77	360	4,675
8	**103 Wautier**	Rowhouse	Residential	170	925	5,441
9	J. de Lalaing	lot	vacant	145	601	4,144
10	**56 Le Titien**	Rowhouse	Residential	61	255	4,180
11	**54 Le Titien**	Rowhouse	Residential	93	1,100	11,827
12	**18 Bourgogne**	Lowrise	Commercial	244	5,000	20,491
13	**76 Mich. Ange**	Rowh./MM	Residential	837	5,050	6,033
14	**17 Palmerston**	Rowhouse	Residential	199	1,500	7,537
15	**19 Palmerston**	Rowhouse	Residential	188	1,300	6,915
16	**21 Palmerston**	Rowhouse	Residential	177	1,000	5,650
17	**23 Palmerston**	Rowhouse	Residential	171	1,410	8,245
18	94 Stévin	Rowhouse	Com./Res.	107	1,400	13,085
1981						
19	24 Arts	Office block	Office	floor	1,675	?
20	**49 C. Martel**	Rowhouse	Residential	55	678	12,327
21	59 Hobbema	Rowhouse	Residential	183	1,550	8,451
22	**48 J.W. Wilson**	Rowhouse	Com./Res.	85	1,050	12,353
23	**41 P. le Bon**	Rowhouse	Residential	73	600	8,219

Source: Répertoire des ventes publiques réalisées pendant les année 1980–1991 par les notaires de l'arrondissement de Bruxelles, 11 vols. (Brussels: Editeur ventes notariales, 1980–92).

[a]The designations Rowh./MM and Rowh./MB describe the so-called *maison de maître* and *maison bourgeoise* respectively. A *maison de maître* is usually a townhouse made of *pierre de France* (high-quality cut limestone) and includes a coach entrance, large reception areas, and lavish interior decorations. A *maison bourgeoise* is a modest variant.

[b]Area in square meters listed by the notarial records refers to the size of the lot and not the total area of lot and building.

[c]The price in thousands of Belgian francs reflects the price paid for land and improvements—usually a building.

[d]The land value listed is not the price of a square meter but rather the ratio of the price paid for the property over the area of the lot.

[e]The properties listed in bold lettering are situated in the core of the quartier and in close proximity to the rues de la Loi and Belliard and the rond point Schuman.

No.	Address	Type bldg. [a]	Land use	Area[b] (m^2)	Bf [c] (000)	Value[d] (m^2)
24	196 Stévin	Rowhouse	Residential	191	1,620	8,481
25	32 de Pascale	Rowhouse	Residential	196	850	4,334
1982						
26	1 Archimède	Office block	Office	?	4,500	?
27	74 Clovis	Rowhouse	Residential	116	430	3,707
28	13 de Pascale	Rowhouse	Residential	109	925	8,486
29	23 de Pascale	Rowhouse	Residential	212	650	3,066
30	33 de Pascale	Rowhouse	Residential	173	650	3,764
31	39 de Pascale	Rowhouse	Residential	242	1,350	5,579
32	41 de Pascale	Rowhouse	Residential	176	1,170	6,648
33	38 J.W. Wilson	Rowhouse	Residential	170	1,150	6,765
34	20 Le Tintoret	Rowhouse	Residential	60	1,200	20,000
35	5–15 Marteau	Lowrise	Industrial	2744	9,200	3,353
36	2/4 Maerlant	Rowhouse	Residential	41		
	6 Maerlant	Rowhouse	Residential	50		
	8 Maerlant	Rowhouse	Residential	124		
	16 Maerlant	Rowhouse	Residential	750	5,000	5,181
1983						
37	20 Arts	Block of Apts	Residential	228	1,025	4,556
38	116 Belliard	House	Residential	223	2,500	11,211
39	118 Belliard	House	Residential	190	1,800	9,474
40	25–7 Charlem.	Lowrise	Com./Hotel	545	25,000	45,865
41	23 de Pascale	Rowhouse	Residential	212	715	3,373
42	33 de Pascale	Rowhouse	Residential	173	715	4,140
43	171 Noyer	Rowhouse	Residential	125	1,600	12,800
44	20/22 Patriot.	Rowhouse	Com./Res.	70	500	7,184
45	2/4 Maerlant	Rowhouse	Residential	41		
	6 Maerlant	Rowhouse	Residential	50		
	8 Maerlant	Rowhouse	Residential	124		
	16 Maerlant	Rowhouse	Residential	750	21,500	22,280
46	79–81 Vérones.	Rowhouse	Res./Wareh.	353	1,000	2,833
47	37 Maelbeek	Rowhouse	Residential	130	650	5,000
1984						
48	120 Belliard	Rowhouse	Commercial	90		
	10 Maerlant	Rowhouse	Commercial	290	8,000	21,053
49	6 Commerce	Rowhouse	Commercial	70	350	5,000
50	41 de Pascale	Rowhouse	Residential	176	1,400	7,955
51	39 de Pascale	Rowhouse	Residential	242	900	3,719
52	112 Joseph II	Rowhouse	Res./ruin	248	3,300	13,306

(Continued)

(TABLE 7, *continued*)

No.	Address	Type bldg. [a]	Land use	Area[b] (m^2)	Bf [c] (000)	Value[d] (m^2)
53	72 Le Corrège	Rowhouse	Residential	175	1,550	8,857
54	5 Livingston	Rowhouse	Res./ruin	170		
	7 Livingston	Rowhouse	Residential	170		
	9 Livingston	Rowhouse	Residential	171	3,000	5,871
55	32 Mich. Ange	Rowhouse	Residential	299	1,850	6,187
56	10 Science	Office block	Office/1st fl.	97	500	5,181
57	10 Science	Office block	Office/4th fl.	97	500	5,181
58	49 Spa	Rowhouse	Residential	138	1,475	10,507
59	53 Stévin	Rowhouse	Residential	143	1,200	8,392
60	55 Stévin	Rowhouse	Residential	223	1,520	6,816
61	63 Stévin	Rowhouse	Residential	197	1,100	5,584
62	19 Toulouse	Rowhouse	Res./Worksh.	304	1,220	4,013
63	53 de Mot	Rowhouse	Residential	70	370	5,286
1985						
64	109 Confédér.	Rowhouse	Residential	141	1,000	7,092
65	74 Cortenber.	Rowhouse	Residential	200	1,640	8,200
66	64 de Pascale	Rowhouse	Res./uninhab.	90	100	1,111
67	116 Joseph II	Rowhouse	Residential	172	800	4,651
68	110/108 Jos. II	Rowh./lot	Residential	392	2,500	6,378
69	15/21 Living.	Rowhouses	Residental	300	3,250	10,833
70	27–29 Living.	Rowhouse	Residential	106	1,750	16,509
71	9 Science	Office block	Office	280	1,850	6,607
72	14 Toulouse	Rowhouse	Residential	107	150	1,400
73	17 Toulouse	Rowhouse	Residential	293	1,350	4,608
1986						
74	41 C. Martel	Rowhouse	Residential	88	1,300	14,808
75	33 Marteau	Rowhouse	Residential	95	750	7,895
76	20 Montoyer	Rowhouse	Com./Res.	145	6,500	44,828
77	40/a Pacific.	Rowhouse	Com./Res.	87	780	8,966
1987						
78	40 Beffroi	Rowhouse	Residential	160	4,500	28,125
79	35 J. Lalaing	Rowhouse	Residential	138	1,725	12,500
80	115/119 Stévin	Rowhouses	Residential	418	7,500	17,943
81	38 Van Ostade	Rowh./MB	Residential	103	1,750	16,990
1988						
82	159 Confédér.	Rowhouse	Residential	67	1,050	15,672
83	78 Cortenberg.	Rowhouse	Residential	125	4,600	36,800
84	117 Froissart	Rowhouse	Com./Res.	199	3,000	15,060
85	5 Jenneval	Rowhouse	Residential	128	3,650	28,516
86	172/4 Jos. II	Rowhouses	Residential	200	6,500	32,500

No.	Address	Type bldg. [a]	Land use	Area[b] (m²)	Bf [c] (000)	Value[d] (m²)
87	176 Joseph II	Rowhouse	Residential	85	7,000	82,353
88	65 Noyer	Rowhouse	Com./Res.	421	7,300	17,340
89	9–13 St. Quent.	Rowhouses	Residential	401	14,000	34,913
90	15–7 St. Quent.	Rowhouses	Com./Res.	304	19,000	62,500
91	111–3 Stévin	Rowhouses	Residential	285	8,300	29,123
92	169 Stévin	Rowhouse	Com./Res.	93	4,600	49,462
1989						
93	65 Stévin	Rowhouse	residential	180	10,500	58,333
94	12 Toulouse	Rowhouse	Com./Res.	101	4,850	48,020
95	38 Trèves	Rowhouse	Com./Res.	264	7,000	26,515
1990						
96	71 Archimèd.	Rowhouse	Residential	310	12,200	39,355
97	5 Belliard	Midrise block	Office	742	139,000	187,332
98	42–4 Confédér.	Rowhouse	Commercial	72	withdrawn	
99	20 Le Corrège	Rowhouse	Residential	190	withdrawn	
100	1 Sq. Marguer.	Apartment	Residential	40	4,501	112,525
101	97–99 Noyer	House	Com./Res.	130	3,005	23,115
102	88 Patriotes	Rowh./MM	Residential	111	withdrawn	
1991						
103	71 Archimèd.	Rowhouse	Residential	310	6,100	19,677
104	183 Belliard	Rowhouse	Residential	125	withdrawn	
105	138 Confédér.	Rowhouse	Com./Res.	83	3,500	42,169
106	3 Pavie	Rowhouse	Residential	124	withdrawn	

estate industry professionals operating in Brussels to indicate on standard maps of the quartier[40] (1) Land value/rent gradient, and (2) directions of CBD expansion within and outside the boundaries of the quartier.[41]

---

40. See footnote below for full reference on the survey. Since the information supplied by the respondents was in most cases proprietary, the results are presented in aggregate form alone.

41. Respondent were asked to indicate the direction(s) of CED expansion by drawing arrows. The symbolization employed ranged from standard arrows marking direction, to arrows of variable length and/or width, and/or arrays of different numbers of arrows indicating intensity of expansion.

The Land Value/Rent Profile

The responses to the first request gave a clear idea of the spatial linkage between the physical expansion of the EC institutions and the valuation of property. There appears to be a very strong coincidence between highly priced real estate and the locations of the EC executive and legislative headquarters. The large, modern buildings facing the avenue des Arts constitute an exception. Their situation places their owners or occupiers roughly equidistant from some of the EC-occupied buildings and the Parliament, the Royal Palace, a number of the federal ministries, and the historic parc de Bruxelles. This set of locations was typically priced at BF 9,000 per square meter (rental price) and above.

Areas of uniform rental value were typically identified as street blocks or sets of street blocks. Interestingly, however, some of the respondents chose to make their characterizations in terms of arrays of façades facing particular streets or avenues. For example, the respondents who marked avenue de Cortenberg as a highly desirable market typically indicated the line of buildings facing the avenue and not the entire street blocks which also face rues Franklin, Stévin, Michèl Ange, and Le Corrège (fig. 20).

There was significant agreement concerning which areas of the quartier command the highest point of the market: properties facing the avenue des Arts, properties surrounding the Schuman roundabout, and properties facing the avenue de Cortenberg between the Schuman roundabout and rue Le Corrège.[42] A quarter to two-fifths of the respondents included in the set of highest demand properties on the square Frère Orban, or at least its northern side, and/or the street blocks bounded by the rues Froissart, Belliard, and Breydel and the avenue d'Auderghem. Fifteen percent also included the square de Meeûs in this set. The Centre International du Congrés and its environs, as well as the Berlaymont, were conspicuously absent from the lists of most: two of twelve respondents included the Berlaymont, and only one marked the rues Wiertz and Vautier and the new CIC multimillion-dollar facility leased by the European Parliament as areas in great demand. Areas typically commanding slightly lower rental prices (BF 8,000–9,000 per square meter), included the bulk of the street blocks between rue du Commerce in the west, rue de Trèves in

---

42. One respondent extended the "top value" array to include the intersection of Cortenberg-Véronèse and Cortenberg-Le Titien.

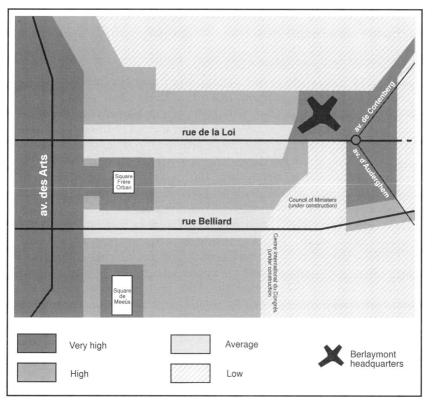

Fig. 20. Distribution of land value in the quartier Européen-Léopold

the east, rue Joseph II to the north, and rue de Luxembourg to the south.

Interestingly, approximately 40 percent of the respondents did not include in this category the buildings lining the rues Loi and Belliard, the main thoroughfares of the quartier. These were indicated as commanding lower prices. With traffic congestion and air and noise pollution reaching new heights by the day, the multilane thoroughfares are clearly deemed less marketable than the quieter and on occasion prettier streets in the rest of the indicated area. Four respondents included the buildings lining the avenue des Nerviens; one extended the area eastward to include the buildings facing the avenue de Tervueren (outside the limits set by the study); one respondent extended it northward to include the place Madou (Olivetti Tower); and one respondent extended it southward to the rue du Trône.

That none of the respondents indicated the remaining portion of the quartier (the northern side of the quartier which is largely residential) as desirable simply suggests that it is of little demand among the clientele of the respondents. That is a little puzzling. A building-by-building survey of the northern part of the quartier, however, reveals that the buildings lining the three fin-de-siècle squares Marguerite, Ambiorix, and Marie-Louise have a significant caché among consumers researching prestigious addresses. A number of diplomatic representations, professional organizations, and institutions have been long-time residents in this overwhelmingly residential portion of the quartier. The attraction is clearly the high aesthetic quality of the parks and waterworks, as well as the retrospective ambiance of lines of old and handsome townhouses.[43]

The side streets of the quartier Nord-Est, regardless of their occasional picturesque quality, fall outside the office market. They constitute a separate market for residential and small-scale commercial space catering to two different clienteles: affluent Europeans advancing northward into the quartier Nord-Est, and long-time Belgian residents and more recently arrived Maghrebin and Turkish residents retreating farther north.

The Direction of CED Expansion

Responses to the request for indications of CED expansion were varied. The responses were very simple: arrows pointing to a specific point in the quartier or an area marked in the quartier or its environs. Pressure for expansion appears to be greatest in the southern and eastern boundaries of the quartier (fig. 21): eastward toward the parc Léopold and the place Jourdan, northeastward along the avenue de Cortenberg, to the cité Linthout and the square Eugène Plasky, east of the parc du Cinquantenaire toward the square de Princesse J. de Mérode, south of the parc du Cinquantenaire to the place St. Pierre, east along the avenue de Tervueren, south to the place de Luxembourg and the rue du Trône, and north of the rue du Marteau

---

43. An extraordinary example of how effective marketing can whip the perspective clientele into a frenzy is the case of the handsome eclectic townhouse at square Ambiorix 17. Purchased in 1988 for BF 5.5 million and meticulously restored at a cost of BF 8 million, the house was bid on two occasions during 1989–90 to BF 40 million. Interview with owner, Mme A. H. [name withheld upon request], Brussels, 8 April 1990.

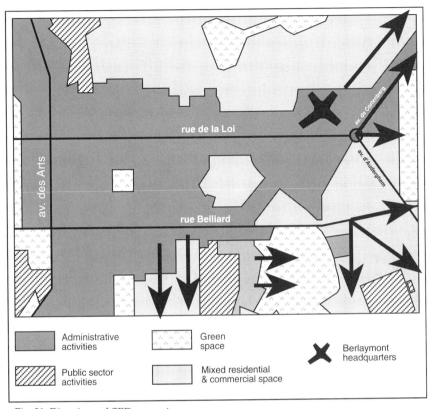

Fig. 21. Directions of CED expansion

into the contiguous residential-commercial areas of the municipality of Saint-Josse.

Certain of these areas are associated with widely publicized controversy: expansions of the Communities or the office sector in general in the quartier Nord-Est are considered intrusive by the population of the quartier, as in the case of the three condemned street blocks to the east of the EC headquarters building Berlaymont, between the boulevard Charlemagne and the rue Livingstone. In this case, the preservationists won. After significant maneuvering, the Belgian state gave up its plans to use these blocks in the expansion of the EC facilities. This was seen as a key decision determining the northern boundary of the EC administrative park (fig. 22).

In response to this decision, however, and because of continuing demand for office space in the quartier, the state and urban developers have turned their attention to the southern flank of the quartier

Fig. 22. Saved derelict rowhouses on rue Joseph II. *Top:* Corner of rues Joseph II and Livingstone. *Bottom:* Corner of rues Joseph II and Taciturn. In the background stands the EU Charlemagne building.

Européen-Léopold: there has been controversy concerning the construction of an extension to the large Commission conference facility Centre Borchette on the place Jourdan, in place of housing and commercial space which had been promised to the community by the developer. This original promise came about as a result of negotiation: negotiation is again guaranteeing the construction of some residential and commercial space in that lot, but the EC will also get its extension built.

Perhaps the most publicized of all cases involving the extension of European space is the erection of the Centre International du Congrés on the site of the demolished Brasserie Léopold. Overshadowing the beaux arts gare du Luxembourg by Santenoy, the neighborhood of the rues Wiertz and Vautier, and the museums on the parc Léopold, the large privately financed conference center is perceived as a forward-placed growth pole which will extend the office sector into yet more residential areas to the south, displacing long-time residents and removing more of the nineteenth-century built landscape. Adding to the controversy is the fact that these residential areas are occupied by a working-class population which will not be likely to withstand the pressure and enticing offers of prospecting urban developers.

A similar situation exists in the northwestern flank of the quartier, along the rue du Marteau. With the continuing development of the street blocks lining the southern side of the rue du Marteau, the residents on the northern side of the street sense that they are on the first line of fire (fig. 23). They have been resisting continuing development by lobbying the city and Regional governments.[44] The activist group in charge of this effort claims that the people occupying the

---

44. Anecdotally, when I was surveying the rue du Marteau, armed with cameras and describing the land use and architecture of the street on my miniature tape recorder, an angry-looking middle-aged woman kept pace with me in her car. In spite of my efforts to ignore her and go about my business, she eventually caught up with me and asked if I was involved with the urban developers who had recently completed the construction of the new Banco di Roma building across the street (fig. 23). When I explained to her the benign purpose of my survey, she invited me into her home on the rue du Marteau and we discussed the situation. She was one of the central figures in the "resistance movement." She gave me a tour of her townhouse and garden and explained to me that the modest façades of the rue du Marteau shelter a long-time, small but vibrant art colony. Indeed, from her garden I could see a number of ateliers in the adjacent houses. This raises the question of subsidiarity: how many townhouses with productive ateliers are worth office buildings? Is the preservation of the rue du Marteau art colony inconsequential to an irresistible trend of postmodernization, or is it the expression par excellence of what is wrong with that process?

Fig. 23. Old and new buildings and land uses on the rue du Marteau. *Left*: The untouched part of the rue du Marteau between the rues des Deux Eglises and Spa. *Right*: The newly constructed building of the Banco di Roma on the south side of the rue du Marteau.

street blocks north of the rue du Marteau—mostly working-class immigrants—will be easily overwhelmed by the urban developers.

The issues of sensibly priced housing, preservation of open space, and mixicity of urban function considered by the *plan de secteur* come into sharp focus in this case, as in the case of the place Jourdan and the gare du Luxembourg. The urban developers appear to resist the concrete territorial definition of urban functions that would produce a *zone sanitaire* around the quartier Européen-Léopold and hamper its physical expansion. The issues are not economic alone. The members of the quartier's growth coalition are politically vulnerable. Its tenure does not depend on the level of satisfaction of urban developers and their clients, but on the general approval of the direction the city is given by its increasing international role.

Evolution of Land Use

Distribution of land uses has changed since the years the quartiers Léopold and Nord-Est were quiescent residential communities dotted with small-scale commercial establishments.[45] These were communities with a stable urban morphological frame and demographic profile. The insertion of the European Communities into the quartier in 1965 with the construction of the Charlemagne building has changed this polka-dot landscape of uses into a wildly diversified quilt of competing land uses which more often than not displace the original uses. Land-use changes at this scale were not an entirely new phenomenon. They were common occurrences during the nineteenth century inside the Pentagon city, when Flemish neighborhoods were being replaced by new planned quartiers or large public-building complexes.

This morphological transformation can be seen as the latest manifestation of an itinerant and changing Brussels central business district: the bulk of increasingly complex business and administrative activities may be seen to have shifted eastward since the foundation of the city a thousand years ago. Dynastic competition shaped the early economy and polity of Brussels. From small-scale commercial activity on the lowland dock yards of the Senne, the maze of specialized markets on its banks, and the Grand Place (simple/unipolar

---

45. A full presentation and analysis of the quartier's urban morphology is presented in chapter 7. This shorter presentation covering only land use addresses strictly the discussion of economic choice and market trends.

structure), business activity evolved during the industrial revolution to include financial services. A process of managed centralization after the formation of the Belgian state created coalitions of political and economic elites. These gave rise to an alternative finance business node away from the industrial river area around the parc de Bruxelles and the new Belgian national institutions (compound/bipolar structure). The political landscape has again changed with federalization and the hosting of the European executive bureaucracy.

Owing to a process of managed complex convergence, which has brought European economies into an ever closer cooperative regime since 1957, business and administrative activities are distributed in Brussels among a number of poles: various sites in the Pentagon, the quartier Nord, the quartier Européen-Léopold, the avenue Louise, a number of suburban locations such as the boulevards du Souverain and de la Woluwe, and a number of extraurban locations such as the municipalities of La Hulpe and Kraainem (complex/multipolar structure) (fig. 24). The center of gravity appears to have crept eastward once more, this time outside the Pentagon city, to the quartier Européen-Léopold. Exceptionally, the quartier Européen-Léopold is emerging as a specialized central business district with clear referents to the European integration enterprise. Administrative and business activities which do not have a direct bearing on that process are generally sorted out because of the high rents in the quartier. This tendency is not however without exceptions. The quartier Européen-Léopold is a prestigious address, and as such attracts all uses that can afford the price of locating there.

The presence of the European Union is dominant throughout the quartier Européen-Léopold, and its influence pervasive in the evolution of circulation patterns, building types, and land use. In very broad terms, the EU is causing the following three patterns of development:

(1) *The emergence of a multitiered circulation system.* The emerging circulation system includes, first, a local traffic network, which corresponds largely to the nineteenth-century street frame and includes underground mass transit traffic; second, a through-put system of high-speed urban highways, connecting the quartier with the existing system of beltways; third, a rail system which will connect Zaventem airport directly to the quartier and will include connections to the existing railroad network at the gare du Luxembourg (under consideration).

(2) *The emergence of a specialized built environment.* The transformation of the built environment includes the gradual removal of nine-

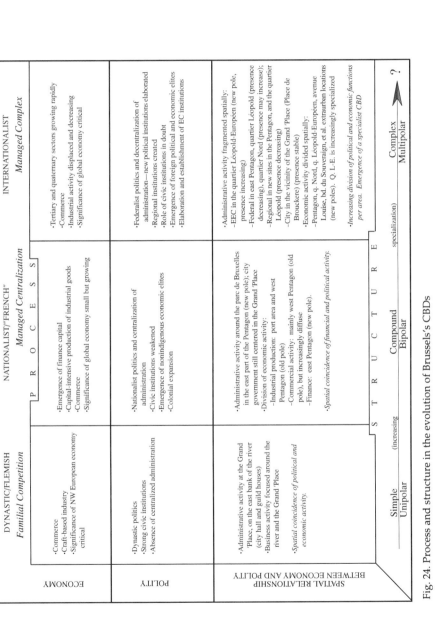

Fig. 24. Process and structure in the evolution of Brussels's CBDs

Fig. 25. Land use 1992: EU and diplomatic missions

Fig. 26. Land use 1992: EU and Belgian government

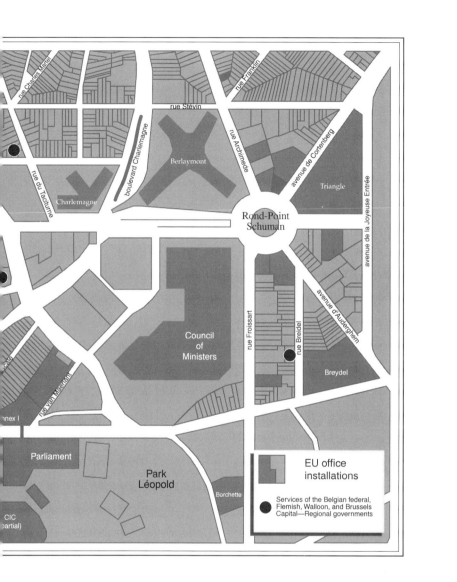

Fig. 27. Land use 1992: EU and EU-related business

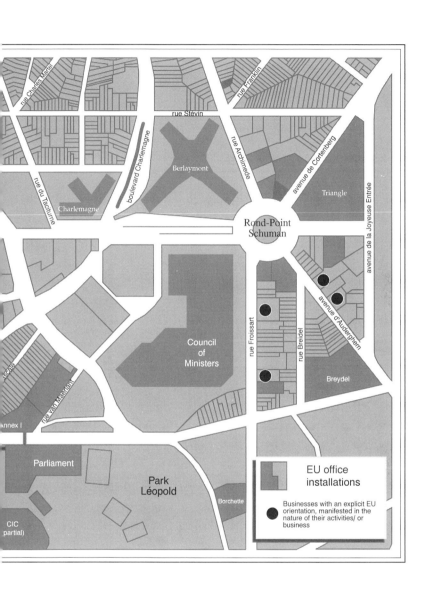

Fig. 28. Land use 1992: EU and other business

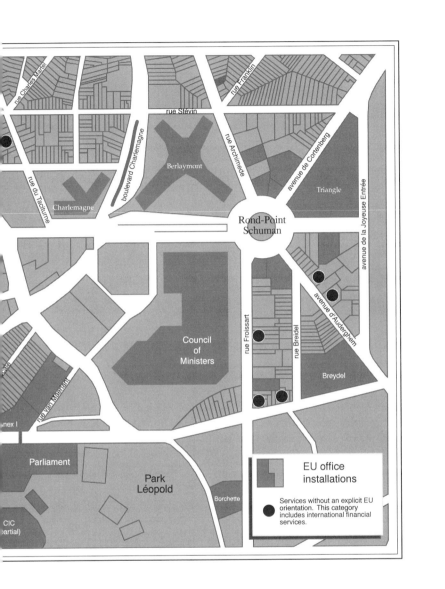

teenth-century buildings, the construction of mid-rise and high-rise buildings, and, as of late, "megaprojects," as well as the reduction of green space and open space.

The valuation of the quartier's land market has had a direct influence on the evolution of building types in the quartier. In turn, developments in EU policy, such as the Single European Market and the continuing integration process under the Treaty of Maastricht, influence the upward valuation of the quartier's land market and consolidate Brussels as a world-class political and business center. The rental clientele of the late 1980s and early 1990s has clearly been more upscale than was that of several decades ago, and higher-quality office accommodations are hence in demand. Characterless office buildings constructed in the late 1950s through the 1970s are in little demand and now vacant; they are slowly being replaced by new, lavishly constructed buildings in granite, brass, and glass. While the visual impact of these changes is noticeable, it is not sufficient to transform the quartier into a showcase of modern architecture and urban planning.

(3) *The emergence of a new cross-section of land uses.* The emerging land uses are vastly different from the residential, small-scale commercial and cultural/recreational land uses for which the quartier was conceived. The new land uses relate to the four broad categories illustrated in figures 25–28:

- European Community administrative activities; European Community diplomatic activities (permanent representations of the member states to the European Communities); other diplomatic activities (foreign missions to Belgium, the EU, and NATO); representation of Regional organizations (for example the European Free Trade Association); representation of industrial sectors and professional groups (for example, the Union of European Employers' Federations UNICE); representation of cultural and religious communities; international press activities
- Belgian federal administrative activities; Regional administrative activities
- Information-intensive professional activities relating to the European Communities (for example, consultants catering to the Single Market); financial services relating to the EU (especially front-office banking and insurance operations); EU research institutions and think tanks
- Information-intensive professional activities without a specific marketing target; financial services (especially front-office bank-

ing and insurance operations); other research institutions and think tanks; specialized service establishments (i.e., courier services); specialized commercial establishments (i.e., computer equipment and office furniture); hotels and residential hotels; eating establishments

In rapid decline or eliminated in the majority of the street blocks in the quartier are residential use (especially owner-occupied rowhouses), small-scale commercial activity (for example, grocery and clothing stores), green/recreational space, and open space, generally defined.

The new land uses are seen to replace the old ones in a very piecemeal way within each street block. With residential and general commercial activities in full retreat, the new blocks are largely monofunctional. Exceptions are the street blocks on either side of the rues Toulouse and Pascale in the heart of the quartier (fig. 29). The preservation of residential uses in these four blocks can be attributed to four factors: first, the location is relatively undesirable, as the railroad traverses the area. Second, the lots are very small; too many lots belonging to different proprietors would have to be purchased and consolidated to accommodate a single office building, which would present logistical difficulties and likely raise total purchase price. Third, these street blocks have become a symbolic battleground for preservation-minded pressure groups and the press. Finally, there is sufficient supply of alternative locations for development to make these politically sensitive street blocks undesirable targets for place entrepreneurs.

One of the key issues in the campaign against the growth of the EU presence in Brussels has revolved around the emergence of monofunctional street blocks as the standard building blocks of a monofunctional quartier. Also criticized is the urgent and haphazard way in which the nineteenth-century built environment has been removed to make way for lackluster office buildings which could easily have been constructed in more recently built areas a short distance away. At the core of this political debate are the very different ways in which different interested parties evaluate the amenities and drawbacks of the quartier.

Advantages and Drawbacks of the Quartier

The cognoscenti of real-estate development speak of their priorities in terms of amenities and drawbacks to them and their clients, as well

Fig. 29. Rues Toulouse and Pascale remain residential.

as land value morphology of the quartier. The questionnaire seeking information on land values/rents and the direction of expansion of the quartier Européen-Léopold also surveyed opinions on the significance of transportation and accessibility, the nature of administrative functions, the level of satisfaction with business conveniences, the comparative cost of land in Brussels and abroad, and the significance of a variety of amenities and drawbacks:[46] the matrix below summarizes the cumulative scores (table 8).[47]

The survey reveals that private developers and real-estate agents share very few of the preoccupations of conservation-minded pressure groups and residents. This is not very surprising, since each constituency has widely different priorities and objectives. As general remarks and assistance to reading the matrix I offer the following observations:

My expectation was that respondents would express an opinion as to which foreign CBD was most attractive to business customers. After all, the price of office space in London and Paris has declined in the last five years, causing clustering in prices in major European office markets. The results of the survey were unequivocal. These firms are investing in Brussels because its office space is extremely well priced. The recession of the early 1990s has accented further this condition. Amsterdam alone is more competitively priced than Brussels, although price by itself is clearly not the most important factor for those interested in the European integration enterprise. Mr. T. Fenwick of Jones Lang Wootton, one of the respondents, maintains that land rents are not necessarily relevant to location decisions, but that costs other than land rents (which he unfortunately did not specify) are more important. He noted that rents are only 10–25 percent higher in the quartier than in other more competitive markets and hence are not an important factor.[48]

Amenities that did not directly affect business activity or relate to

---

46. Survey in English and French entitled "Questionaire for Brussels-Based Real Estate Development Firms and Independent Real Estate Agents: Examining Market Forces and Patterns in the Brussels CBD and Its Environs," organized by author, 1991–92. The survey solicited the opinions of realtors about the so-called *Espace Bruxelles-Europe* area, which largely falls within the more generously bounded quartier Européen-Léopold.

47. The firms were asked to supply basic information about their structure (type of corporation, range of activities, primary market focus, size (in Brussels and worldwide, if applicable), and location of corporate headquarters.

48. T. Fenwick, Jones Lang Wootton, Handwritten note on completed survey, 1992; in author's possession.

TABLE 8. Survey of Brussels-based real-estate development firms and independent real-estate agents

	Scores in percentage points[a]: Significant — Not significant			

Transportation

1.1. Its general centrality in the Brussels Agglomeration	75	25	0	0
1.2. Its good access to rail stations	25	25	50	0
1.3. Its good access to Zaventem Airport	0	75	25	0
1.4. Its good access to the beltways	50	25	25	0
1.5. Its good access to other business nodes in the city	25	50	25	0

Administrative functions

2.1. Its proximity to the EU institutions	80	20	0	0
2.2. Its proximity to the Cité Administrative	0	0	50	50
2.3. Its proximity to the Parliament, Palace, and ministries	0	50	30	20

Business conveniences

3.1. Its offering business support services considered				
3.1.1. Adequate	60			
3.1.2. Good	40			
3.1.3. Excellent	0			
3.2. Its offering better business support services than				
3.2.1. The Pentagon city	90			
3.2.2. The quartier Nord	75			
3.2.3. Avenue Louise	60			
3.2.4. Avenue de Tervueren	60			
3.2.5. Boulevard du Souverain	40			
3.3. Its aggregating like business uses	25	50	0	25
3.4. Its density of EC-related firms	75	25	0	0
3.5. Its offering ready accessibility to EC-level officials	60	40	0	0

Land rents

4.1. Its favorable land rents compared to CBDs abroad				
4.1.1. London (various locations)	100	0	0	0
4.1.2. Paris (center)	100	0	0	0
4.1.2.1. (La Défense)	100	0	0	0
4.1.3. Amsterdam	25	65	10	0
4.1.4. Frankfurt	70	30	0	0
4.1.5. Berlin	100	0	0	0
4.2. Its favorable land rents compared to other nodes				
4.2.1. The Pentagon city	0	0	15	85
4.2.2. The quartier Nord	20	10	40	30
4.2.3. Avenue Louise	25	20	40	15
4.2.4. Avenue de Tervueren	25	10	20	45
4.2.5. Boulevard du Souverain	20	0	45	35

[a]Ratings read left to right: Great significance – average – small – not significant.

	Scores in percentage points[a]:	Significant — Not significant		

Amenities

5.1.	Its proximity to good restaurants, bars, and cafés	0	40	50	10
5.2.	Its proximity to the green amenities of the				
	5.2.1. Parc Léopold	0	0	10	90
	5.2.2. Cinquantenaire	0	0	15	85
	5.2.3. The squares Frère Orban	0	0	20	80
	5.2.4. Marguerite	0	0	0	100
	5.2.5. Ambiorix	0	0	10	90
	5.2.6. Marie-Louise	0	0	0	100
5.3.	Its proximity to the cultural amenities of				
	5.3.1. The Ilôt Sacré and the Pentagon city	0	0	0	100
	5.3.2. Art nouveau architecture in the squares	0	0	0	100
	5.3.3. The parc du Cinquantenaire museums	0	0	0	100
	5.3.4. The parc Léopold Natural History Museum	0	0	0	100
5.4.	Its proximity to high-quality housing in				
	5.4.1. High-rise apartment buildings	0	0	0	100
	5.4.2. Mid-rise apartment buildings	0	0	20	80
	5.4.3. Row/townhouses	0	0	50	50
	5.4.4. Single-family homes	0	0	50	50
5.5.	Its proximity to good schools	20	30	10	40

Drawbacks of Espace Bruxelles-Europe

6.1.	Government regulation (urban planning)	80	20	0	0
6.2.	Public opposition to office sector	25	50	25	0
6.3.	Quality of construction	20	35	25	20
6.4.	Age of buildings	0	25	75	0
6.5.	Scarcity of renovated/redeveloped space	0	40	50	10
6.6.	Inappropriate building types	20	30	50	0
6.7.	Absence of security personnel	0	20	55	25
6.8.	Lack of amenities in buildings	0	30	40	30
6.9.	Inadequate parking facilities	30	50	20	0
6.10.	Inadequate hotel accommodations	30	50	20	0
6.11.	Level of land rents	0	30	70	0
6.12.	Lack of architectural distinction in buildings	75	25	0	0
6.13.	Lack of a grand planning vision	50	40	10	0
6.14.	Lack of shopping opportunities	25	25	25	25
6.15.	Inadequate housing	0	0	30	70
6.16.	Inadequate schools	0	20	10	70
6.17.	Traffic congestion	0	60	10	30
6.18.	Crime	0	0	10	90
6.19.	Proximity to low-income quartiers (North)	0	0	20	80
6.20.	Elimination of green spaces	50	25	10	15

the European Communities fared poorly among all respondents: proximity to the good restaurants, bars, and cafés lining the rues Archimède, Franklin, and Stévin near the Schuman roundabout left the respondents rather indifferent.

The parks Léopold and Cinquantenaire and the squares Frère Orban, Marguerite, Ambiorix, and Marie-Louise were consistently deemed of little or no importance to locational decisions, although Belgian firms gave them consistently higher ratings than did non-Belgian firms. Even so, only 25 percent of Belgian firms rated the parks Léopold and Cinquantenaire and the square Frère Orban as of average importance; none deemed them of great importance. Real-estate developers and their clients rarely provide for the preservation of green space in the interior of street blocks, a standard feature of nineteenth-century vernacular urbanism in the city of Brussels.

All respondents agreed in assigning little importance to the proximity of the quartier to the cultural amenities of the Ilôt Sacré and the Pentagon, the art nouveau architecture in the squares, and the various museums within walking distance. This position may be seen as contradictory to the developers' complaints about the lack of architectural vision in the quartier, which will be discussed in the next section.

In general terms, proximity to high-quality housing was of greater importance: The preferred housing types were townhouses, single-family homes, and mid-rise apartment buildings, with high-rise apartment buildings of least importance. Yet none of the respondents acknowledged access to good housing stock as an important amenity, although they registered greater interest in good schools and especially in high-quality nursery care: the European Union will soon be moving the nursing service from a townhouse on avenue Palmerston to a substantial new multistory facility on boulevard Clovis.[49]

The respondents were asked to rate twenty drawbacks and externalities of the quartier. The first ten items reflected likely concerns of developers. The remaining items on the list are concerns of resident action committees and pressure groups, such as the Atelier de Recherche et d'Action Urbaines and Inter-Environnement Bruxelles. The survey did not mark these two sets of items in any special way, so as not to prejudice the respondents.

The responses highlighted the deep divisions between the con-

---

49. As the demographics of the quartier illustrate, the denizens of the EU institutions usually live outside the quartier in the affluent and greener communes east and southeast of the Pentagon.

stituencies: First, of the ten drawbacks of highest importance to ARAU and IEB, seven were rated by the urban development firms as of little or no significance. There was, however, strong agreement between development and pressure groups on questions concerning the lack of architectural distinction in buildings and the lack of a grand planning vision. Traffic congestion—a drawback of the quartier and the city which receives enormous press—was rated of average significance. Second, the respondents themselves could be rated for their lack of civic-mindedness. Approximately 30 percent of the respondents rated *all* of the items in the first set as of no significance. Third, nationality of the firm appears to bear some significance in the pattern of response: British firms were consistently less concerned than their Belgian counterparts with the first set of items.

The concern of urban developers with the lack of architectural distinction in the new buildings of the quartier and the lack of a grand planning vision for the executive seat of the European Union is noteworthy. While the beautifully landscaped nineteenth-century and early twentieth-century parks and squares, museums, and historically significant architecture of the quartier Européen-Léopold were rated as amenities of little or no significance to the locational choices of the clients of the respondents, the *lack* of such aesthetic amenities was identified by 50–75 percent of the respondents as an important drawback of the quartier. This leads one to question why urban developers and the city's and region's planning authorities have not seized on the *pragmatic* value of such readily available aesthetic amenities in the quartier to promote "architectural distinction in building" or "a grand planning vision." Façadisme is occasionally exercised around the city with varying success to placate pressure groups and planning regulators but is perhaps shortchanging these aesthetic resources, as well as derailing any possibility of creating an architectural ensemble worthy of the seat of the European executive branch (fig. 30).

Certain pressure groups for architectural preservation would much rather have the European enterprise cease to expand in the quartier Européen-Léopold and move outside the bounds of the nineteenth-century city altogether (i.e., to the vacant gare Josaphat rail yard) in an effort to salvage the standing nineteenth-century urban fabric.[50]

---

50. Press release of Hervé Cnudde: "Le projet ARAU de construction d'un second pôle Européen sur la gare Josaphat pris en compte par les pouvoirs publics," Brussels, 19 September 1991; "Un projet culturel pour l'Europe" designed in 1982 for ARAU by architects Brigitte d'Helft and Anne Gérard of the Ecole d'Architecture pour la Recon-

From the perspective of the urban developers, however, the connection has yet to be made between the competitiveness of the quartier Européen-Léopold as a land market and the quartier's aesthetic resources.

Of the first set of drawbacks two items are of great significance to the attempts of pressure groups to influence the shape of the quartier Européen-Léopold: availability of housing and shopping, and the "mixicity" of the quartier. The "Declaration de Bruxelles" signed by urban activists and planners in 1978 spells out this concern:

> All intervention in the European city should focus on what have always made up cities: roads, squares, avenues, street blocks, gardens . . . or simply "quartiers" [neighborhoods].
>
> All intervention in the European city should ban city highways, single function neighborhoods, and token green spaces.
>
> We must not promote "industrial," "commercial," or "pedestrian" zones but only neighborhoods which would include all the functions of urban life.[51]

Pressure groups and the liberal press have voiced their support of expanded housing and shopping in the city in general and the quartier in particular, coming in direct conflict with the clear development trend creating a monofunctional office sector in the quartier Européen-Léopold. Lack of housing was rated as of no or little importance by 100 percent of the respondents. Lack of shopping was rated evenly across the spectrum of choice. Regrettably, the survey did not distinguish between different kinds of commercial activity which could have spoken to the issue of mixicity. It is uncertain how the respondents interpreted the term "shopping."

Of the second set of items, the specter of government regulation was deemed a drawback of great significance by 80 percent of the respondents. Government qua planning regulations has historically been a perceived rather than a real threat to private-sector urban development. The very stratification of planning standards (*plan de secteur* and *plans particuliers d'aménagement*) and the unplanned look of the post–World War II Brussels built landscape illustrate the light-

---

struction de la Ville.

51. "Déclaration de Bruxelles," *Archives d'Architecture Moderne* 15 (1978), in *La reconstruction de Bruxelles: Recueil de projets publiés dans la Revue des Archives d'Architecture Moderne de 1977 à 1982* (Brussels: Editions des Archives d'Architecture Moderne, 1982), p. 31.

Fig. 30. Façadisme on the rue de l'Ecuver in the Pentagon

handed manner in which the public sector has related to the needs of business.

Public opposition to the growth of the office sector in the quartier is a phenomenon of less concern: 25 percent (mostly Belgian firms) identified it as of great importance, while 50 percent considered it of average importance. At a time of increased public awareness concerning the need to preserve the artistic and architectural heritage of the city to attract and retain residents, the vagueness of the current planning regulations can be utilized to oppose the megaprojects of the private sector. This can occur, however, only if new political leadership of the city and the Brussels-Capital Region dissolve the regimes of cooperation that currently propel development activities in the quartier (see chapter 6).

Prominent nonproblems included the level of land rents, which were deemed quite competitive within the city and across other land markets, the absence of security personnel and the lack of amenities in buildings. All these were rated by at least 70 percent of the respondents as of little or no importance.

Inadequate parking facilities received more attention, as did inadequate hotel accommodations. Indeed, on-street parking is nearly impossible during the business day, and while special parking garages abound for EU officials and non-EU executives, the general public and visitors to the quartier are at a disadvantage. Thirty percent of the respondents rated inadequate parking as an important drawback, while 50 percent rated it as of average importance. Shortages in hotel accommodations were deemed significant at least to some degree by 80 percent of respondents.

These data evoke Logan and Molotch's hierarchy of "place entrepreneurs." All the respondents are brokers of real estate and related services and have a concrete financial stake in the health of the quartier's land market. Depending on the size of their portfolios and their access to the levers of government, they can be considered either "active entrepreneurs," or "structural speculators." They should be considered a part of the Brussels "growth coalition."

In general terms, urban developers and independent real-estate agents responding to the survey appear, first, to utilize simple locational logic in assessing the value of the centrality of the quartier within the Brussels metropolitan area; second, to emphasize the key importance of its price competitiveness when compared with other major European land markets; third, to recognize the significance of amenities that enhance productivity; fourth, to disregard existing cultural and aesthetic resources; fifth, to protest the lack of architectural distinction and planning vision in the quartier; and, finally, to emphasize the negative impact of government intervention in the land market.

One extremely important contradiction is patent in the responses: On the one hand, the *presence* of government as regulator of urban development is viewed in negative terms, and on the other hand, the *absence* of government or alternatively some strong, guiding hand in conceiving a grand planning vision for the quartier is also seen in negative terms.

The phrase "grand planning vision" describes the imagined morphology and management of a large piece of territory (for example, a quartier) over an extended period of time. Whether it is an individual or a collective entity that decides what this vision may be is not of primary importance here. Certain groups like ARAU and IEB jockey for the leading role in determining the direction to be taken. The respondents seek leadership in urban planning as long as such planning will not compromise the short-term profitability of the quartier. This vacillation, I believe, is partly responsible for the unimaginative and destructive nature of urban development of the quartier.

Given that the European Communities appear to be consolidating their facilities at the very heart of the quartier Européen-Léopold between rue Stévin (north), avenue de Cortenberg and avenue d'Auderghem (east), rue d'Arlon and the place de Luxembourg (west), and the place Jourdan (south), around a renovated Berlaymont, the new Council of Ministers building, and the Centre International du Congrés, their significance in the market of the quartier is clear: they are the center of gravity for an estimated hundred thousand persons residing in Brussels and its environs who are connected in some way to EU activities, including EU executives and their families, some seven thousand lobbyists, approximately seventeen hundred foreign firms with a foothold on the quartier Européen-Léopold, a growing number of diplomatic services of the 113 states with envoys to the EU, and the majority of the international press representations.[52] The trend of centralizing EU-related activities in Brussels is well illustrated by the decision to relocate both the revived Western European Union (the exclusively European defense body) and Eurocontrol in

---

52. *The European Public Affairs Directory 1992* (Brussels: Landmarks, S.A., 1991); Thiry, p. 169; and Staes, *Europe at Stake.*

the quartier.[53] The allure is the profit that can be made by building a "Manhattan sur Senne." The success of this endeavor will be measured by the extent to which the quartier Européen-Léopold growth coalition will be able to weather the vagaries of the market and produce a professional and lay consensus on what the capital of a uniting Europe should look like.

53. M. J. Bamber, ed. "Property Report Belgium," *Richard Ellis Research* (September 1992): 1. Eurocontrol is the organization responsible for monitoring and predicting air traffic flows. It may provide the basis for a simple European air traffic administration.

6

National, Ethnic, or City Interests First? The Quartier as Political Artifact

I N THE SEARCH for the persons, structures, and processes that have shaped the quartier Européen-Léopold, political factors must also be taken into account. The practice of politics in Brussels is no less extravagant than the façades of its Grand Place guild houses, and no less elaborate than a saraband reflective of its the Spanish past. Réné Schoonbrodt, president of the Atelier de Recherche et d'Action Urbaines and Inter-Environnement Bruxelles, two leading public pressure groups with special interest in Brussels urban planning, suggests that every Belgian's political profile can be identified along four spectra: the political left versus the political right, Catholic versus secular, royalist versus republican, and federalist versus unitarian.[1] The possible combinations of these cultural-political expressions define the range of partisan politics and the great number of coalition governments that have ruled Belgium—and Brussels—since World War II.

While some of these critical political forces originate in the international and national political arenas, since the constitutional reform of the state in 1989, it appears that Regional, cultural community, and local interests and politics are taking on primary importance. These forces have produced political structures by which urban policy is decided and carried out in Brussels. These structures include the national government, the Regional government as an element of the 1989 federal scheme, the corporate linkages between major financial players in the Brussels market, and the numerous government/ private-sector regimes of cooperation which have emerged since 1957. The missions of these different political structures are qualified

---

1. Réné Schoonbrodt, interview, Brussels, 10 September 1991.

to some extent, in the case of Brussels, by the special status of the city as the provisional seat of the executive of the European Communities.

What makes the Brussels political network so difficult to penetrate, understand, and describe, is that *all* the formal political structures that define modern Belgium are represented in the city's politics. Brussels is no longer just a focusing lens for Regional and cultural identity; it is also the cultural exception that confirms the rule of Regional separation that has otherwise marked Belgian politics. It is the coveted prize of the Walloon and Flemish communities, and has also served as an arena for business experimentation by private financial interests. Some of these political structures are permanent, such as the constitutional ones, and some are relatively ephemeral ones, such as the contractual agreements between private real-estate developers—but what makes them all work is a small constellation of personages that has adorned national politics and economics since World War II. The structures cannot produce results without the intervention of these personages, and the personages cannot achieve their goals—whether self-serving or noble—outside the context of the structures (fig. 31). Hence the fundamental theoretical question we pose in every chapter: who possesses the power to transform the quartier? Is it institutions like the Regional Council and Executive Branch that orchestrate the cooperation regimes? Is it the firms doing business in the quartier, which act in a predictable manner as economic actors in a market environment? Or is it the individual actors, such as Jean-Louis Thys, the secretary of state for the Brussels-Capital Region and minister for public works, or the captains of the Belgian construction industry Armand Blaton and Charlie De Pauw who mold the forces thanks to their economic might?

The Brussels political process can be described as outrageous but not exceptional. The guiding hand of the Brussels land market is certainly not invisible if one believes the accusations of corruption of the system by bribery, nepotism, strong-arm tactics, and downright undemocratic practices in certain political and financial circles in Brussels and Paris. These accusations have been made both by a group of liberal journalists and by members of the European Parliament and are serious enough to have become important in the understanding of the quartier as an artifact of political forces, some of which are residual while others are new.[2] Moreover, the investigation of the

---

2. Georges Timmerman, *Main basse sur Bruxelles: Argent, pouvoir, et béton* (Brussels: Editions EPO, 1991).

political structures and cooperation regimes that have brought certain parties and institutions together will allow us to assess the relative significance of these forces for the market trends described in the previous chapter and for the aesthetic trends discussed in chapter 8.

The Constitutional Reform

An agreement reached in May 1988 by the political parties belonging to the eighth Wilfred Martens coalition government became the basis for the special law of 12 January 1989 which established the status of Brussels in the new federal structure and today regulates the city's governmental institutions.[3] This agreement was brokered after eighteen years of negotiation, legislative obstructions, and planned delays between the two cultural communities (the Flemish and the Walloons), and after a considerable measure of political acrobatics by those who wanted it most, namely the conservative Social Christian Party (PSC/CVP). The status of Brussels as symbol and economic asset has clearly been of great significance for both linguistic communities. The long negotiation process assured that all parties concerned would leave the table with some of their needs satisfied. This is the nature of politics around the world, but it takes a new meaning in Belgium.

The constitutional reform has played a significant role in realigning the political forces that have traditionally exercised influence over Brussels affairs either officially or covertly. The federalization of the Belgian state and the elevation of the Brussels metropolitan area, to the status of "Region" with rights, privileges, and obligations that are roughly equal to those of the other two regions, signify the relative diminution of the national government's influence, and a commensurate increase in the influence of local politicians. As the role of the national government has been deemphasized, so has the role of the nineteen municipal councils comprising the Brussels metropolitan area, which were deemed archaic and inappropriate vehicles for policy.[4] The long time required for Brussels to attain Regional status illustrates the fact that this institutional and power realignment was a

---

3. *Moniteur belge* (14 January 1989). The *Moniteur belge* is the official government publication which announces new legislation, executive orders, and royal decrees.

4. C. Vigneron-Zwetkoff, "L'administration des grandes agglomérations belges," *Administration Public* (1983): 142.

Structure

- Political institutions • Constitution • Planning regulations
- Firms (i.e., strategy, style) • World economy (i.e., business cycles, securitization of real-estate investment, growth of tertiary, quaternary sectors)
- Pressure groups

Agency

- Politicians • Bureaucrats/technocrats • Lobbyists • Reporters • Activists
- Entrepreneurs • Securities traders • Voters • Quartier residents

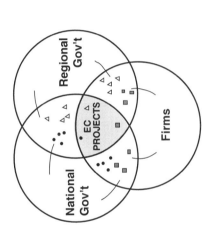

Key officials from the national and regional governments, working through their institutions, plot and manage European Community installations with the assistance of certain business personalities.

Fig. 31. *Top:* Weighing the influence of structure and agency in Brussels. *Bottom:* Institutions and elite individuals at work.

very contentious issue. It may also illustrate why Brussels urbanism has been plagued with problems.[5]

While there were relatively few conceptual difficulties to resolve in positioning the regions of Flanders and Wallonia in the federal scheme, it was very difficult to define, or even justify, what Regional status would mean for Brussels, and for the two competing cultural communities. The Flemish community was clearly on the defensive. Brussels was considered Flemish cultural territory, but was thoroughly infiltrated by Walloon/French culture and population. This concern was underlined by the conviction that an alliance of Brussels with either of the other two regions could tip the federal balance against the solitary Region, which was likely to be Flanders.[6] Since the Walloons would never have accepted the absorption of Brussels into Flanders, the idea of allowing Brussels separate Regional status would ensure that Flanders would be forced to tolerate a Walloon-dominated, French culture enclave in the middle of Flemish Brabant. This was unacceptable to the Flemings.[7]

Up to that point in time, Brussels was the capital of a unitary state and lacked any special privileges, but rather had a number of constraints imposed upon it by the national government owing to its role as capital. For example, the national government was responsible for

---

5. The first instance where the Regional and federal schemes appeared in reference to the status of Brussels was during the 1930s. The spreading influence of French-culture Brussels on its Flemish surroundings was noted by the Centre d'études pour la réforme de l'état, a private think tank. "It is primarily the expansion of the Brussels agglomeration and its French influence on Flanders which has persuaded Flemish circles to demand a reform of the State based on a federal scheme"; translated from *La réforme de l'Etat* (Brussels: Centre d'études pour la réforme de l'état, 1937), p. 389. In the 1960s, this notion came to be known as the "oil slick of Brussels." The origin of the constitutional law of 12 January 1989, however, can be traced to Article 107 *quater* introduced by Parliament. This article created—at that time in name only—three regions: Flanders, Wallonia, and the Brussels-Capital Region. The language used was appropriately vague and allowed for a wide variety of options for the Brussels-Capital Region. It made it clear however that the regions would not be simple administrative or executive vehicles but full-fledged political institutions. P. Wigny, *La troisième révision de la Constitution* (Brussels: Bruylant, 1972), p. 179.

6. Philippe De Bruyker, "Bruxelles dans la réforme de l'Etat," *Courrier hébdomadaire du CRISP* [Centre de recherche et d'information socio-politique], 1230–31 (1989): 11.

7. The Flemish community has been divided on which institutional approach to the Brussels Region question served Flemish interests best. They want to protect Flanders against the influence of Brussels, while grabbing the opportunity to "reconquer" Brussels. Yet another point of view is that today Flanders is both demographically and economically stronger that Wallonia, and conceding to the federal scheme would simply signify a lost opportunity to control the helm of the entire country. Wigny, p. 37.

key urban planning decisions, such as the building of new train stations, the maintenance of public transportation, and the establishment of international organizations (EC and NATO) in city territory. The national government sought to ensure that the national interest would be advanced. Its decisions could override decisions of the city council that were deemed contradictory to the goals set for the city by the national government.

The second step toward Regionalization was taken by the second coalition government of Leo Tindemans. On 1 August 1974 the Parliament created the provisional Regional institutions which were to usher in the full regionalization of Brussels with status equal to that of Flanders and Wallonia.[8]

During eight years of inaction (1980–88), Brussels was governed by a ministerial committee consisting of a presiding minister and two secretaries of state, one of which had to belong to a linguistic group other than that of the minister. These personages were drawn from the slate of national politicians serving in the national Parliament. There were no representatives of the Brussels population in this executive body, which may be the reason urban decisions during this time did not reflect the needs of Brussels and the Brussels Region, but rather those of the national government. A report by the Centre d'études des institutions politiques holds this interim arrangement responsible for the physical and economic deterioration of the city during the early 1980s. Financial problems stemming from the unresolved status of the city paralyzed important consultative and planning bodies such as the Regional Economic Council for the Brabant, stalled environmental reform, urban planning, and land management efforts, and aggravated problems relating to the presence of significant numbers of resident aliens (North Africans, Turks, and non-Belgian Europeans). The only regulatory regimes put into place were a series of royal decrees on Brussels housing.[9]

Apart from discussion of the shape of the Brussels Regional institutions, one additional critical issue remained unresolved in 1988: the

---

8. Full text of the legislation appeared in the *Moniteur belge* (22 August 1974). The political agreements of July 1978, known as the "pacte d'Egmont" and "accords du Stuyvenberg" produced legislation implementing the 1970 Article 170 *quater*. The opposition from Flemish groups was so strong that the fourth coalition government of Tindemans fell on 11 October 1978. H. Lemaître, *Les gouvernements belges de 1968 à 1980: Processus de crise* (Brussels: Chauveheid, 1982), p. 211.

9. E. De Schrevel, ed., *Un Statut pour Bruxelles* (Brussels: Centre d'études des institutions politiques, 1988), p. 19.

assignment of a cultural identity to the Brussels Region, and the political capital it would represent for the Flemish and Walloon communities. If Flanders was the stronghold of Flemish culture and the
Dutch language in Belgium, and Wallonia was the stronghold of
Walloon culture and the French language, what was to be the official
cultural stamp of Brussels? Should the Flemish allow French culture
to swallow the rest of the Brussels Region? Should the Walloons concede cultural parity to the Flemish, when they, the Walloons, were
the overwhelming linguistic majority in the city?[10] These concerns
led to the recognition of a tricultural, trilingual community system in
the country (Flemish-Dutch, Walloon-French, and German), with
Flanders, Wallonia, and certain districts of the Liège province as territorial expressions of each linguistic group. This was deemed necessary because there was not complete spatial coincidence of the cultural groups with the three regions created by the constitutional
reform.[11] The unequal representation of the cultural communities in
the Brussels Region necessitated provisions for the protection of Flemish interests.[12]

The solution, codified in Article 59 *bis* of the Constitution, con-
sciously avoided considering the inhabitants of the Brussels Region
as a separate cultural community, distinct from those of Flanders and
Wallonia.[13] The two dominant cultural communities were instead
considered as cohabiting the Brussels Region. The cultural communities were allowed linguistic jurisdiction only within their respective
territories: Flanders and Wallonia. They could therefore have no exclusive territorial claims within the Brussels Region. The cultural

---

10. And if this was not complicated enough, what about the rights and needs of the
German-speaking community of southeast Belgium which requires representation in
the capital? In this study, we will limit ourselves to the issues dividing the Flemish and
the French cultural communities.

11. This system is not fail-safe either. For example, if the French-speaking community is made up of all French-speakers in the Walloon and Brussels regions, what of the
3–5 percent of French-speakers living in the Flemish provinces? According to the
Constitution they do not belong legally to any community at all! Fitzmaurice, p. 111.

12. The Flemish community has waged a great effort throughout this century for fair
linguistic representation in the education system and the national cultural institutions.
Belgian ethnoregional division and strife expressed through the competition for linguistic rights is explored in Murphy, *The Regional Dynamics of Language Differentiation in
Belgium*.

13. It is difficult but not unthinkable to define postindependence Brussels culture as
distinct from its Flemish and Walloon parent cultures. Moreover, with 25 percent of its
population being foreign-born, one could argue that Brussels is as representative of
Flemish or Walloon culture as New York is of American culture.

communities could exercise jurisdiction only over specific institutions conceived to belong exclusively to one or the other group.[14] This arrangement served as a cultural tie between the Dutch-speaking communities of Brussels and Flanders and assured the continuing existence of a Flemish presence in Brussels, which the Regional arrangement could not guarantee.[15]

A Special Status for Brussels: Legislating the
Brussels-Capital Region

The implementation of Article 59 *bis* concerning the role of the cultural communities in Brussels paved the way for settling the Brussels-Capital Region constitutional crisis. In May 1988 the government decided that the political climate was suitable to push for a solution to the question of the Regional status of Brussels. Furthermore, if Brussels were to aspire to the status of capital of the new united Europe of 1992, it appeared absurd and embarrassing for it to be situated in an institutional no-man's-land.

What complicated the negotiating of the Regional power structure was the existence of a metropolitan level of government, which encompassed the executive and legislative branches of the nineteen constituent municipalities of Brussels. This fragmented metropolitan area authority was the legacy of the early nineteenth century, when the municipalities surrounding the still modest but growing Pentagon city strove to secure an appreciable amount of authority in city affairs, and succeeded. Therefore, any attempt to recast city management in terms of federal regions had to accommodate the particularistic interests of the nineteen municipalities that make up the Brussels metropolitan area, and by consequence, the political interests of the municipal-level politicians who sometimes constitute notable political power bases.

Modifications to the earlier relevant articles of the Constitution opened the way for "the empowerment of the Brussels-Capital Region in the areas of provision of services to [Flemish] persons living outside an area [Flanders] although culturally linked to it" (*matières*

---

14. For example, the Flemish Theater Institute clearly falls under the jurisdiction of the Flemish community.

15. A. Méan, "La casse-tête des matières personalisable à Bruxelles," *Belgique: la révision constitutionnelle et la régionalisation; Problèmes économiques, politiques, et sociaux* (Brussels: La Documentation Française, 1981).

personnalisable),[16] for provision of services relating to the management of the metropolitan area,[17] and for specification of the cultural
matters for which the cultural communities of Brussels and the
Regional executive would to be responsible.[18] The process consolidated functions and responsibilities, which had thus far rested with
the central state or with the metropolitan area of Brussels, at the
Regional level (fig. 32).

After intense negotiation the special law of 12 January 1989 was
adopted, creating the institutions to govern the Brussels-Capital
Region. The result was an institutional structure with a "variable geometry," which allowed members of the governing institutions to
hold multiple offices. They could simultaneously hold offices at the
level of the Region, the metropolitan area, and the cultural community.

The Brussels-Capital Region is limited to the nineteen bilingual
municipalities of the old Brussels metropolitan area, a limitation approved of by the Flemish constituency, which has traditionally feared
the territorial expansion of Brussels. The special law invested the
Region with the following institutions and powers (fig. 33):

- Legislative Council (Conseil de la Region de Bruxelles-Capital)
  of seventy-five members, with five-year terms of office, directly
  elected by the population of the Region. Ethnolinguistic balance
  is of the essence. Candidates are slated either for the Flemish-
  speaking or the French-speaking portion of the council seats.
- Executive of five drawn from and elected by the council. Ethnolinguistic balance is again a critical ingredient: the executive
  should include at least two councilors of each linguistic group.
  Furthermore, three secretaries of state, of which at least one
  must be of the minority linguistic group, *assist* the executive,
  though they are not members of the executive. Sessions of the
  executive are also attended by representatives of the cultural
  communities. The executive assigns its six "portfolios" to its
  members, as in government (Finance, Economic Policy and Employment, Housing, Public Services, Urban Planning, and Local
  Government).[19]
- Authority to initiate bills (*ordonnances*) in the council, and issue
  decrees (*arrêtés*) by the executive for matters concerning portfo-

---

16. Constitution, Article 59 *bis*, paragraph 4 *bis*.
17. Ibid., Article 108 *ter*, paragraph 2.
18. Ibid., paragraph 3.
19. Fitzmaurice, p. 123.

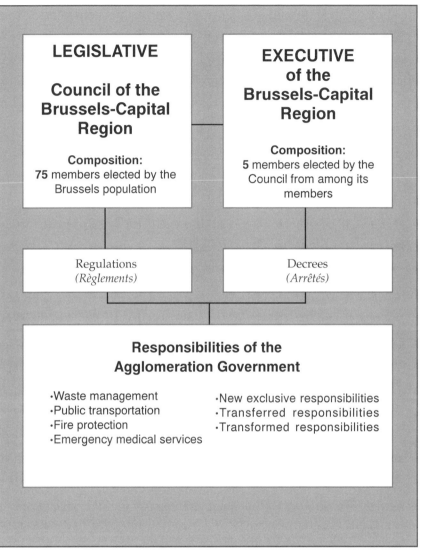

LEGISLATIVE

Council of the Brussels-Capital Region

Composition:
75 members elected by the Brussels population

EXECUTIVE of the Brussels-Capital Region

Composition:
5 members elected by the Council from among its members

Regulations
(*Règlements*)

Decrees
(*Arrêtés*)

Responsibilities of the Agglomeration Government

·Waste management
·Public transportation
·Fire protection
·Emergency medical services

·New exclusive responsibilities
·Transferred responsibilities
·Transformed responsibilities

Fig. 32. The government of the Brussels agglomeration. Adapted from Serge Loumaye, *Les nouvelles institutions bruxelloises* (Brussels: Courrier hebdomadaire du Centre de recherche et d'information socio-politiques, no. 1232–33, 1989), p. 31.

lio issues, as well as matters of local government (such as fire fighting), which had previously been the responsibility of the Brussels metropolitan area authority. Laws of the national government have precedence over the decrees of the Region, and potential conflicts are resolved by arbitration at the Council of State (Conseil d'Etat).[20] This hierarchy of legal authority known as *tutelle*, or administrative supervision, allows the national government to intervene in Brussels when national interest is threatened: a case in point is the agenda of the central government to make Brussels the capital of the European Community regardless of the wishes of the inhabitants of the Brussels-Capital Region.

- Authority and means to raise funds, seek grants from the national government, and disburse funds for Brussels projects initiated by the Regional authorities. [21]

The implications of this major restructuring of governance in Brussels is further confused by the fact that the elected candidates for the first Regional council, and therefore executive branch, who are still in power, were drawn from the ranks of the national parliament. According to evidence presented later in this chapter, the patronal linkages of this group to the national government appear to be of utmost importance for the direction of urban planning and land management in Brussels. Under the present composition of the Brussels Regional executive and legislative branches, there is less a devolution of national power to the Region than a reshuffling of personages with links to both national and Regional levels of government.

What importance does this process of regionalization have for the understanding of the political forces shaping the quartier? First, it codifies the shift in the balance of power from the national to the Regional level. For as long as the Brussels Regional government is drawn from the roster of national politicians—as is the current one— the effects of this shift remain contained. Owing to the new legal requirement, the Regional government will be drawn truly democratically and independently of the national government nomenclature, and Brussels affairs will in the future be more or less emancipated from direct national control despite the *tutelle*. This is perhaps why the current government of the Brussels-Capital Region is so keen on finalizing deals with the private sector that will likely persuade the

---

20. *Document parlementaire*, Sénat (1988–89), 514/1, p. 88; and 514/2, p. 118.
21. Outline of institutions largely following De Bruyker, pp. 31–54.

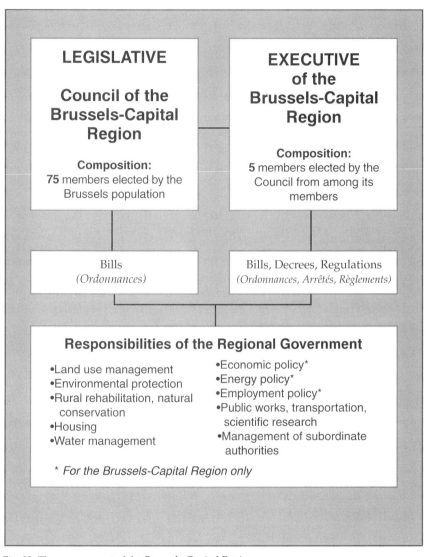

LEGISLATIVE

Council of the Brussels-Capital Region

Composition:
75 members elected by the Brussels population

EXECUTIVE of the Brussels-Capital Region

Composition:
5 members elected by the Council from among its members

Bills
(Ordonnances)

Bills, Decrees, Regulations
(Ordonnances, Arrêtés, Règlements)

Responsibilities of the Regional Government

- Land use management
- Environmental protection
- Rural rehabilitation, natural conservation
- Housing
- Water management

- Economic policy*
- Energy policy*
- Employment policy*
- Public works, transportation, scientific research
- Management of subordinate authorities

* For the Brussels-Capital Region only

Fig. 33. The government of the Brussels-Capital Region

European Communities to stay in Brussels: such deals include the decision to build huge administrative complexes for the Council of Ministers and the European Parliament.

Second, the regionalization of Brussels opens opportunities for smaller political players. Whereas in a unitary state access to and personal relationships with national-level politicians were available only to the social and financial elite of Brussels and Belgium, in the new Regional political environment local relationships and public opinion will be of greater importance. Pressure groups, the Brussels press, and residents' committees can all throw Regional politicians out of office, thus disrupting the traditional regimes of cooperation between government and the private sector that have engendered the EC-centered CED of the quartier Européen-Léopold.

Third, the constitutional reform is important in separating the EC issue from the cultural rivalry between the Flemish and the Walloons over Brussels. Quite surprisingly, both cultural communities appear content with their representation in the new Brussels Regional power structure. Had this issue remained unresolved, the Flemish community would have seen the "Eurocratization" of Brussels as an intolerable challenge to its efforts to salvage Flemish culture in that city. Today the building of Europe in Brussels is viewed as a culturally value-free issue. Local concerns over its urban impact cut across cultural trenches.[22]

Fourth, the credibility and electoral health of the conservative Social Christian Party (PSC/CVP) rest on the success of the Brussels European venture. The new Regional arrangement changes the familiar rules of engagement by which the national government has been promoting the European Communities in Brussels since the signing of the Treaties of Rome. The risk of disruption of this process by regionalization, and hence electoral failure, is great. The EC building program has suddenly become time-sensitive. Coalition with the private sector is now the vehicle that ensures the completion of the program.

Fifth, as far as Brussels urban planning and land management are concerned, while it has been difficult to translate these patronal linkages between business and the national government into positive policy for the Brussels Region under the unitary state, it is at last easier to envision the formation of a coherent urban planning code and vision under the Regional system. As we discussed in the previous chapter, the Specific Urban Management Plans (*plans particuliers d'aménagement*: PPA) have repeatedly been used by government/ private-sector coalitions to redraw neighborhoods and parts of neighborhoods in favor of extending the territory dedicated to EC activities. We can expect that PPAs will be closely monitored under the Regional arrangement.

The European Dimension of the National and the Regional Agendas

The national government would maintain that Brussels won the European "Sweepstakes" in 1957 by attracting the World Exposition and the executive branch of the newly founded European Communities. The monetary rewards stemming from the establishment of the Commission and the Council of Ministers have been significant for the economy of the city. The building of Europe in Brussels has been a perpetual Olympics for the coffers of the national government and the city's entrepreneurs.

Has this European dimension of Brussels has been put to good use by authorities planning the Brussels of the future? What are the considerations that drive the national and Regional agendas transforming the city and the quartier Européen-Léopold into the political and urbanistic nexus of the new Europe? With what standards and urban planning parameters is the task carried out? The optimist would ask: is it a visionary effort replete with symbolism—a twenty-first-century European version of L'Enfant's Washington? Is it the showpiece of European urbanistic ingenuity and technological innovation that will set new standards for modern cities everywhere? Is it a daring experiment in central executive district design that will serve as a model for New York, London, and Tokyo? The pessimist would ask: how much planning is at the base of these urban morphological transformations? Do the current transformations follow the questionable urban traditions that destroyed Victor Horta's art nouveau gem Maison du Peuple, and planned the "Houstonscape" of the Cité Administrative? Is greed and profit maximization the objective of a small group of firms and politicians possessing privileged information to the detriment of any social and community amenities, and is this the group with the greatest impact on urban development?

The endorsement of one or more of the above perspectives on the transformation of the quartier must depend partly on official rhetoric and partly on physical evidence. While the physical evidence gives incontrovertible testimony to certain aspects of the transformation, such as the changing land use, building types, intensification of economic activity, and changing land value gradient, an examination of official rhetoric offers information on the national and Regional governments' agendas and anxieties and helps to identify issues that are politically sensitive.

Rhetoric about the installation of the EC facilities in Brussels exemplifies what is politically palatable to the voters of the Brussels-Capital Region. The authorities seek to downplay fears about the impact of the growth of the EC institutions on the availability and pricing of housing in and around the capital, and the emergence of monofunctional neighborhoods, and especially office districts which would be depopulated after business hours. Possible fallout of such a transformation would include an increase in crime in these neighborhoods. They are also concerned about minimizing the impact from the marginalization of small businesses in the quartier Européen-Léopold. Finally, environmental externalities, such as auto emissions and noise pollution, the explosion of automobile circulation, increasing congestion in the quartier Européen-Léopold (which also interferes with circulation citywide), and the diminution of inner-city green space, are quickly becoming political issues important to the Brussels electorate, as is the removal of the nineteenth-century and fin-de-siècle buildings which define the character of the quartiers Léopold and Nord-Est.

These concerns have been voiced over a number of years by interest groups such as the Atelier de Recherche et d'Action Urbaines and Inter-Environnement Bruxelles. While educating the Brussels population about these issues, ARAU, IEB, and like organizations have become quite powerful in detecting urban planning violations and in intervening by rallying the liberal press. Their judgments have often been well researched, thoughtful, and reasonable, and they have slowly moved from being policemen and teachers to leaders of a significant portion of public opinion.

The authorities have employed subtle publicity in seeking to placate such interest groups and to respond to the mounting concern of the Brussels population. This effort focuses on the ways in which the above concerns are actually integrated into planning for the European Communities, and of course, on the positive monetary implications of their presence in Brussels. The culmination of this publicity effort was the report *Espace Bruxelles-Europe* produced jointly by the Regional government, the conservative urban affairs think tank CERAU, and Professor José Vandevoorde of the Free University of Brussels.[23]

The *Espace Bruxelles-Europe (Brussels-Europe Area)* project brings together the urban planning means by which the Regional and national governments aspire to keep the European Communities in Brussels. The rhetoric in the introduction to the sixth and final phase of the report, written by Professor Vandevoorde, acknowledges the problems troubling local organizations. He makes the following remark about the role of Brussels as the capital of the EC:

> A very important fact is the clear desire of the European administrations, the Commission, the Council of Ministers, and the Parliament, to reorganize and regroup themselves in closer quarters, at least by institution. The direction of this process of establishment in the last ten years demonstrates this desire for greater proximity, which is not negligible and whose consequences on the urban fabric should therefore be taken into account.[24]

Article 77 of the European Coal and Steel Community Treaty, Article 216 of the European Community Treaty, and Article 189 of the Euratom Treaty simply state that "[t]he seat of the institutions of the Community shall be determined by common accord of the Governments of the Member States."[25] Since these treaties were signed, various judgments of the European Court of Justice and a number of European Parliament resolutions refer to Brussels, Luxembourg, and Strasbourg—listed alphabetically—as the places where the Community institutions are to conduct their business until such

23. CERAU, *Espace Bruxelles-Europe: Phase F rapport final* (Brussels: CERAU pour le Secretariat d'état à la région Bruxelloise, Administration de l'urbanisme et de l'aménagement du territoire, 1987).

24. "Un phénomène de première importance est la volonté très nette des administrations européennes, commission des communautés européennes, Conseil des Ministres et Parlement, de se regrouper spatialement tout au moins par entité. L'évolution des implantations depuis 10 ans montre cette volonté de rapprochement; aspiration non négligeable et dont les conséquences sur le tissu urbain doivent être prises en compte." CERAU, *Espace Bruxelles-Europe*, p. 6.

25. *Treaties Establishing the European Communities: Treaties Amending These Treaties; Single European Act* (Luxembourg: Office for Official Publications of the European Communities, 1987), pp. 99, 405, 732.

time as a clear decision is made. While Brussels has been the most favored of these locations in being made host of the EC executive branch—the Commission and the Council of Ministers—since the signing of the Treaties of Rome in 1957, as well as certain services of the European Parliament, Community law does not guarantee the continuation of this advantage. The Belgian government has therefore been catering to the needs of the European Communities since the mid-1960s by acting as host and reasonable landlord and by sponsoring the building of appropriate facilities and adjusting the city plan as required, in the hope that the historical balance, convenience, and continuous deposits to the Eurocrat "favor bank" will eventually induce the European decision makers to make Brussels the permanent EC capital. Critics would describe this as the European tail that wags the Belgian dog, with Brussels paying the price. The more suspicious among them would also say that the government of the Brussels-Capital Region is not acting independently of the national government, and again, that Brussels pays the price.

At this time the national government would like to see the provisional status of the city as the seat of these institutions lifted and the European Parliament move permanently to Brussels. This would not be a small feat, as French President François Mitterand made Strasbourg's claim to the European Parliament his personal crusade. The challenge to the French government has nevertheless been mounted. The gauntlet is the vast International Congress Center (Centre International du Congrés) now close to completion by private interests with the tacit agreement of the national and Regional governments. The stakes are high enough and the position of the Belgian government precarious enough for European Community officials to use them as leverage to obtain what they want. During the autumn of 1991, at the height of the controversy concerning the future of the asbestos-contaminated Berlaymont headquarters, and in the face of the Belgian government's decision to renovate rather than demolish and replace it with an enormous new headquarters building, certain high-ranking German officials of the EC presented Bonn as a possible site for EC headquarters, arguing that it was not desirable to preserve the outdated building. Hints about the great availability of prime office space in the old West German capital must have sent Regional and national officials scrambling to avert a public relations disaster.[26]

---

26. Personal interview, Rudolph Schneider, Directorate General XXIII: Enterprise Policy, Cooperatives, and Tourism, Brussels, 13 February 1992.

The decision of Prime Minister Wilfred Martens of the Christelijke Volkspartii (CVP), a moderate partisan of a unified Belgium and royalist, to run in the next parliamentary election in the district of Brussels-Halle-Vilvoorde underlined the desires of the dominant partisan coalition, of King Baudouin and the Court. Marten's task has been to defend the European aspirations for Brussels, while preserving Brussels as a symbol of unity of the two main cultural communities and protecting the unity of the Kingdom.[27] Importantly, in spite of the terrible loss the country sustained by the loss of King Baudouin, King Albert II appears as determined to continue his brother's politique of communitarian unity.

The central concern, however, is whether the problems of Brussels as a modern metropolis under great pressure to cater to needs of the local population can be accommodated in a grand-scale urbanistic strategy which serves a separate national political agenda. The model and the empirical evidence offered below question the wisdom of this approach. The national strategy for Brussels serves certain purposes well—namely attracting the Union and international services to Brussels—but does not necessarily well serve the needs of the city as it moves toward the twenty-first century.

---

27. *De Standaard* [newspaper], 21 June 1991, in Timmerman, p. 7.

7

From the Boudoir to the Trenches: Two "Rational" Urban Games

S
OCIAL AND behavioral scientists have long been exploring the causal mechanisms behind human decision making and have tried to identify a basic unit of analysis that could shed light, on the one hand, on the wider process of human interrelations, and on the other hand, on particular social phenomena which may not appear to be immediately related to individual action. The transformation of the quartier Européen-Léopold is a case in point: physical and social transformations there may rightfully be attributed—at least to some degree—to peculiar modes of decision making undertaken by a variety of agents. Among all the projects undertaken in the quartier, the nearly multimillion-ecu building Centre Internationale du Congrés designed by D. Bontinck/CDG/SV Structure/M. Vandenbossche is the most telling example of how different agents engineer consensus for urban change. The question remains, however, of how "rational" the ensuing quartier landscape may be for all user parties.

For Jon Elster and the rational-choice theoretical establishment, the elementary unit of social life is the individual human action[1]—a notion that may appear odd outside the Western and Westernized cultural realm, but which is basically appropriate for describing social interaction in late twentieth-century Brussels. By "individual," in this case, Elster would include both physical and corporate persons, such as firms and government authorities, thus avoiding questions that structuralism raises about collective and complex entities that influence social organization and interaction, including the behavior of individuals. Still, however, the state-of-the-art thinking

---

1. Jon Elster, *Nuts and Bolts for the Social Sciences* (Cambridge: Cambridge University Press, 1989; reprinted 1993), p. 13.

on rational choice is far from pedantic or rigid. Indeed, the field is frequently dismissed—especially by the left—as a foster child of economic neoclassicism and instrumental rationality, abstracting personal experience fruitlessly within a "structureless" context.

In reality, as Elster notes, "[w]e care about rationality because we want to be rational and want to know what rationality requires us to do."[2] That statement would probably resonate with every utility-maximizing market player in the Brussels quartier, but what Elster would suggest is that it would also resonate with the conservation-minded retiree or bohemian who may be driven by emotions such as indignation, resentment, and revenge. For example, he has illustrated through extensive experimentation that a person may be willing to take a loss (therefore, not maximize utility) rather than be unfairly exploited.[3] So rational-choice theory has also extended into a consideration of the importance of social norms and institutions, as well as emotions, and concedes that there are inconsistencies and inconclusiveness in how humans may actually select one course of action over another.

What one discovers quickly while reading in the rational-choice field is that the conclusions are far from normative and deductive and are based on significant empirical work. The construction of the various "games" that work as technical tools for explaining behavior are not based on a fundamental hypothesis that all rational human beings are utility-maximizing, but rather act as interpretive media that explain—among other things—why there is not general chaos in society, but rather widespread, seemingly paradoxical instances where human beings make compromises, cooperate, and may be willing to undercut their utility. Again, what constitutes a fairly standard path of assault on rational-choice gaming is the notion of "social norms," which according to Elster do not foster outcome-oriented actions. Cynics or absolutists, however, may interpret social norms as "tools of manipulations, used to dress up self-interest in more acceptable garb."[4] Although we can list types of social norms (etiquette, sense of equality, vengeance), it is difficult to operationalize them and integrate them into modeling. Social norms end up being a black box that holds the explanations for why rationality fails.

---

2. Elster, "Some Unresolved Problems," p. 189.
3. Ibid., pp. 185–86.
4. Ibid., p. 118. Elster does not agree with this notion.

The basic and certainly most famous of the rational-choice games is the so-called Prisoner's Dilemma. It involves the assessment of costs and benefits of cooperation or noncooperation by two persons who have been arrested for allegedly committing a crime together. Each prisoner is interrogated separately and has no way of knowing what his accomplice is testifying about. The game suggests that if both cooperate and deny the charges, they are released. If they both divulge their involvement in the crime and incriminate their accomplice, that is, "defect," each receives a modest penalty. If one cooperates with the police and confesses his crime but the other denies it, then the former receives a large payoff and the latter a harsh sentence (the so-called "sucker's payoff"). The game's purpose is to explore how persons make choices which involve variably valuated risks and benefits. The dilemma for the prisoner is whether to risk a harsh sentence for the possibility of going free by seeking to cooperate to keep secret his involvement in the crime.

Rational-choice theory, however, does not always succeed in explaining human behavior and is thus less of a predictive than a prescriptive tool: it is indeterminate when it fails to tell people what to do; and people may act irrationally by failing or simply not choosing to act in accordance with what it tells them to do.

Elster explains the sources of indeterminacy—the logic of backward induction and decision making under uncertainty—by exploring the Hangman's Paradox: a four-round game between two players who are each allowed to disburse the benefits or costs of each round when they know in advance the relative cost and benefit that each chance to cash in represents for each of them (fig. 34). Assuming first that both players are rational and both know that their antagonist is also rational; second, that the playing of all four rounds yields the higher *overall* payoff for both; and third, that the intervening rounds favor significantly one or the other of the players; rationality would dictate that the players cooperate to see the game through to the last round to cash in a high overall reward, although not the highest one each may earn by defecting at one of the intervening rounds. In fact, one player is more likely to defect on the first round in which she is rewarded to avoid the expected "sucker's payoff" during the playing of the subsequent round, when the second player is rewarded at a higher rate.

As Elster points out, and as in the case of the Prisoner's Dilemma game, individual rationality as exemplified by the players' thought exercise in the Hangman's Paradox leads to collective suboptimal-

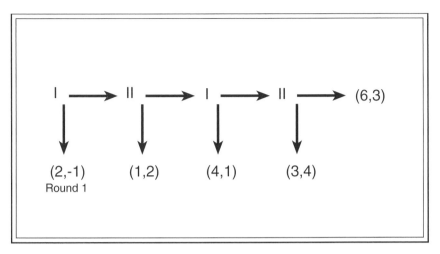

Fig. 34. The Hangman's Paradox

ity.[5] Moreover, the example, he says, has no rationally prescribed behavior. The reason behind the paradox is that self-interest is only one of the motivations that drives individual choice.

Of special interest to us would be the instances where emotions such as "myopia," regret, indignation, and revenge are involved in decisions that at first appear irrational, but in fact are based on the subjective valuation of alternatives by individual players. In this context, "myopia," for example, is the antithesis of foresight, or the ability to be motivated by long-term consequences of action.[6] For example, one has a myopic outlook in organizing his life, if one chooses welfare in the present over welfare in the future. Elster reports that there is good evidence that people act this way with negative consequences on their long-term well-being. The problem with labeling such shortsightedness as irrational is that it may be a preference—albeit objectionable to some—in which the individual chooses to spend all early and leave nothing to his children. Or alternatively, the individual may not have the ability to visualize the lifelong accounting of his welfare. In both cases, and where the schedule for high consumption of one's assets is concerned, once again, rationality does not imply maximization.[7] Elster points out that it is difficult to call myopia irrational or rational, and it is important to take it and

5. Ibid., pp. 181–82.
6. Elster, *Nuts and Bolts*, p. 42.
7. Elster, "Some Unresolved Problems," p. 185.

such elements of the subjective valuation of rewards into consideration.

Regret, indignation, and revenge describe a cast of emotions
which taint rational behavior and produce outcomes that appear at
first to be irrational. As a game where indignation and revenge rule
suggests, people are willing to trade utility for fairness, even when
the payoff is zero (fig. 35).

The first player has the choice of disbursing a reward of 2 to each
or allowing the second player to disburse the rewards of the second
round, which favors the first player and undercuts the benefit to the
second player, that is, benefits 3 and 1 respectively. Knowing that
the alternative move available to the second player will yield zero to
both, the first player assumes rationally that the second player would
prefer for himself a payoff of 1 to zero, since the conceptual basis of
rationality for the paradigm is maximization of utility. In fact, Elster
reports, systematic experimentation has yielded the following three
findings: "First, most people placed in the position of the first player
propose a much more equal division. Secondly, when a very unequal division is proposed by the first player, the second player often
decides to take nothing. Thirdly, most proposals that are made and
accepted are somewhat biased in favor of the proposer."[8]

Assume now the case of community resistance to financial rewards
offered by urban developers which are valuated by the community
as significantly inferior to the utility already offered by the neighborhood in its present state. The indignant responses of the Ateliers
Mommens's residents suggest just such a thought exercise: the
absorption of the threatened street blocks into the developed area
represents a cash reward for the residents, who would be able to
purchase housing with greater standard amenities within or outside
the quartier; the residents, however, valuate their nineteenth-century
built environment more highly than the cash award, and furthermore, are motivated by a sense of fairness in their decision to protect
the immigrant populations residing in the contiguous blocks behind
the Ateliers Mommens.

Robert Axelrod informs us of the conditions under which cooperation will emerge in a world of egoists without central authority.
Again here, by employing a certain degree of license, we can use the
particulars of rational-choice strategies to produce possible game-
theoretic interpretations of relationships among actors in the quartier.

8. Ibid., p. 186.

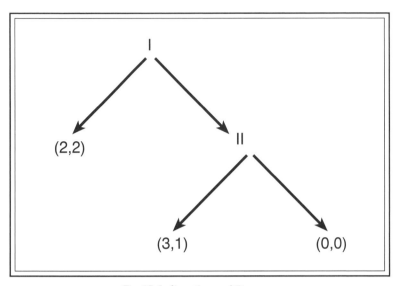

Fig. 35. Indignation and Revenge

Axelrod poses the question of what is the best strategy to ensure cooperation among self-interested parties and starts by echoing Elster's caveat that there is no best rule independent of the environment in which the strategy will be played out.[9] He suggests, however, that in the playing of an indefinite number of games by the same set of players, cooperation can indeed emerge. He points as evidence of such cooperation to the so-called folkways or the elaborate set of norms and etiquette that have emerged in the U.S. Senate: "Among the most important of these is the norm of reciprocity, a folkway which involves helping out a colleague and getting repaid in kind."[10] This norm is not out of place in the Brussels political theater—as it is not out of place in any parliamentary political setting—and can also be extended to describe the regimes between government elites and urban developers as suggested by Georges Timmerman later in the study.

Axelrod concludes that among the strategies ALL DEFECT, ALL COOPERATE, TIT FOR TAT, and the rule "cooperate until the other defects, and then always defect," TIT FOR TAT is the best strategy one can use to assure cooperation. In brief, TIT FOR TAT requires

9. Robert Axelrod, "The Emergence of Cooperation among Egoists," *American Political Science Review* 75 (1981): 309.

10. Ibid., p. 307.

that one cooperate in the first round of a pairwise game and subsequently employ the same stance as the other player. The incentive then is for the second player to cooperate, since she cannot benefit by defecting inside a Prisoner's Dilemma setting, especially when the temptation to defect is eroded to a known high degree with each subsequent defection. In simple terms, TIT FOR TAT makes cooperation the most beneficial response by far in a multiple-game setting.[11] Ideally, the players employ the "nice rule," according to which one commits to not defecting first.[12]

What is useful to us is the concept of a field of isolated egoists, qua homeowners, who fail to cooperate with one another to avert an "invasion" by organized egoists with a common agenda, qua urban developers. Indeed the evidence from the quartier's development suggests that there the residents tended not to band together, the Ateliers Mommens case being an exception. In terms of Axelrod's visualization then, residents employ an ALL DEFECT strategy with respect to each other; since they cannot be enticed to cooperate with each other and instead pursue utility on the basis of their individual sensibilities. ALL DEFECT, Axelrod suggests, is always a collectively stable strategy, since the community can resist invasion by anyone using any *other* strategy, provided that the newcomers arrive one at a time.[13] Again, to use the quartier as an example, if a single urban developer attempts an invasion and is rebuffed by all, then the invasion fails. If, however, the invaders arrive en masse, or even in clusters, then ALL DEFECT as a defensive strategy fails easily, since the invaders can easily establish cooperative TIT FOR TAT, that is, reciprocal regimes between themselves to perform the desired tasks. As Axelrod notes, they will have a chance to thrive. In the case of our quartier, not only would the developers be able to cooperate with each other in creating a marketing intensity—which could erode the entrenched resolve of the residents—but they would also be able to establish a firm foothold in the quartier by purchasing properties which belong either to the state or to individuals who do not reside in the quartier and do not rely on use-value and emotion in assessing

---

11. Ibid., pp. 308–9.

12. One can easily visualize the application of such game-theoretic thinking in nuclear deterrence and the elaboration of a "No First Strike" strategy. The assumption here, again, is that multiple strikes are possible, as indeed is the case. Submarines armed with nuclear weapons are essentially invulnerable to a first nuclear strike and can inflict a punishing second round on the violator of the regime.

13. Axelrod, "Emergence of Cooperation," p. 315.

their attachment to their property. Moreover, the invaders—when acting in concert—may be able to rally on their side external agents who can influence the shape of the quartier: local and national politicians.

The application of strategies to real-world situations reveals the limitations of the theory. Ultimately, the quartier environment is made up by such a tremendous complex of actual or potential interrelationships that the exploration of individual decision making based on modeled pairwise or even n-person games becomes at best an *indicator* of possible outcomes and not a *predictor*—just as Elster warns. Still, I think that rational choice is a good baseline theory for sorting out the social complexes in the quartier.

It is important for geographers to explore this interface of social and behavioral sciences, as political scientists, economists, and sociologists have done. This exploration should not be seen as a direct descendant of neoclassical locational theory, the focus of which was the elucidation of reasoned spatial geometries—"rational" landscapes sorted out of a seeming chaos of social-material driftwood.[14] The phenomenology of neoclassical economic geography as spatial science focused on the regularity, nested geometry, and elegance of patterns produced by capitalist production, while it took for granted—and quickly dispensed with—social complexity as interference.[15] The proposed focus puts the emphasis on the social structure of location rather than on the mere geometry of the pattern. In a manner, I seek to wrench rational gaming conceptions out of neoclassical economic geography and employ them in a process that creates a new type of "place"—the Brussels CED—that is dependent neither solely on a faceless market, nor on the particularistic self-interest of a handful of players. Although full understanding of the quartier's *paysage raisonné* may still elude us, we will have made sense of a sociospatial process in which power relations among actors matter in the silencing, control, modification, and appropriation of an urban landscape.

Among geographers, Trevor Barnes and Eric Sheppard have written on rational agency in space and place. They approach the rational-choice paradigm cautioning against a methodologically individualist theory of society, but also building on Elster's discussion of

---

14. August Lösch's studies of the relationship of space and economy and Walter Christaller's formulation of central-place theory are paradigmatic of such thinking.

15. Trevor Barnes critiques eloquently the deterministic character of neoclassical economic geography in "*Homo Economicus:* Physical Metaphors and Universal Models in Economic Geography," *Canadian Geographer* 31 (1987): 354–95.

decision making under uncertainty. They attempt to demonstrate "that the complex interdependencies of a spatially extensive and interlinked capitalist economy make it extremely difficult for individuals to predict all the consequences of their own and others' actions."[16] They depart further from Elster, who, according to them, "locates the cause [for myopic rational decision making] . . . in individual beliefs."[17] True to what I read as a structuralist paradigm, they attribute the nature of individual action to social relations. Their most important contribution in the cited paper, however, involves their treatment of rational choice in the context of collective action. More specifically, they point to emotions and morals (resembling Elster's "social norms," which, however, are related to one's sense of place) as central to decision making when community interests are at stake—a point missed by Elster, who would always look to individual action divorced from place-specific attachments.

Therefore, for Barnes and Sheppard, "[t]he solution is to include the communities within which people live and work as an integral element in the theoretical analysis of the formation of consciousness, rather than treating this as a contingent modifying factor deployed only in empirical work."[18] They proceed by crediting feelings of guilt and injustice, long-term social relationships, and the routine of everyday life *in particular places* as building blocks for consensus. In other words, they suggest that the key lies "in whether the identities that form in places reflect social relations that privilege collective solidarity or egotism."[19] The contributions here are many: not only do they remind us of the methodological tension between individual agency and structure, but they also inject into the debate the issues of "placeness," geographic scale, and spatial linkages of that action within an interdependent world economy.

The issue of community is brilliantly treated by Michael McIntyre in his study of the paradoxical decision of the Chambonnais to openly shelter Jews during the Vichy period and Nazi occupation in spite of the direct threat of wholesale massacre of the villagers by German Nazi units. Ultimately, moral persuasion by key figures, divisions of opinion toward the "Final Solution" among German commanders, the construction of a community ethic, and "luck" were central to the

---

16. Trevor Barnes and Eric Sheppard, "The Rational Actor in Space and Place: A Re-Evaluation of the Rational Choice Paradigm," manuscript, 1993, p. 16.

17. Ibid.

18. Ibid., p. 28.

19. Ibid., p. 31.

paradox's working in "rational" terms. In McIntyre's words,

> Luck comes in two forms. The Chambonnais were lucky that LeForestier's testimony moved Major Schmehling to exert himself to spare the village. That sort of luck was extrinsic to their project. They were also lucky that they were the sort of people who could carry out such a project. The qualities required were uncommon: courage, discretion, patience, trust, faith, efficiency, tenderness, strength. Before the fact, the Chambonnais did not know that they possessed or would develop such resources. This intrinsic luck, I think, is the key to our moral intuition about the Chambonnais.[20]

Indeed the parable of the Chambonnais self-sacrifice would corroborate Barnes and Sheppard's suggestion that it is the sense of place that makes individuals select "second best" outcomes in rational choice, consequentialist, and utilitarian terms. McIntyre suggests that rational-choice theory in the Chambonnais case has fallen short of the promised goal of full explanation. While it offered a view into the thinking exercises of the Chambonnais before and after they decided to shelter Jewish refugees, it could not have predicted that Le Chambon would shelter refugees. He continues, "[t]hat fundamental decision cannot be understood apart from [the community leader's] moral leadership, the Huguenot tradition of resistance to state authority, and religious character formation, categories that rational-choice theory cannot accommodate."[21]

In spite of the eloquently argued deficiencies of the model, the significance of place and community for understanding individual and collective action in the quartier Européen-Léopold should be explored. If we assume that Barnes, Sheppard, and McIntyre are right, it would be key to determine whether the residents of the quartier have indeed formed a sense of place and an ethical community which would make them check their egotism in favor of community preservation.[22] In the two-person Boudoir Game of attrition, where community spirit does not exist and pecuniary, instrumental interests are dominant, a real-estate corporation (C) and a *rentier* resident (R) will interact in the following fashion (fig. 36): the firm has the choice of either seeking engagement in the quartier with *rentiers* who hold

---

20. Michael McIntyre, "Le Chambonnais Paradox," Polity (forthcoming).

21. Ibid.

22. Gaming choices and valuation of outcomes for the two simulations are based on interviews of quartier residents.

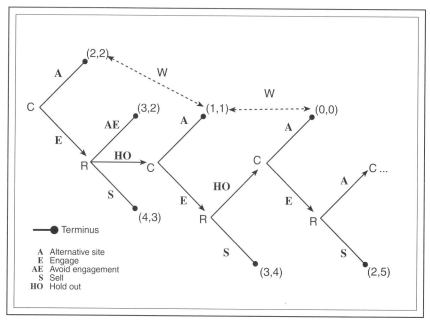

Fig. 36. Boudoir Game: consequentialist-instrumental reasoning

desirable properties, or seeking an alternative site for development.

While the latter choice is a terminus representing a medium-level utility in the form of continuing use-value for the *rentier* and second-best centrality for the firm, the former choice is preferred by the firm because two of the three outcomes represent higher utility.

In the second round, the profit-maximizing *rentier* has three options: first, he can sell the property, securing for himself a reward of 3, and thus extract greater utility than his use-value reward of 2. In this terminal outcome, the firm secures the highest reward of 4 with the minimum transaction costs. Second, the *rentier* may decide to avoid engagement, and therefore not sell. The reward will be equal to his use-value utility, or 2. The corporation is still capable of engaging other *rentiers* in the quartier's land market, although some transaction costs have already accrued, reducing potential utility to only 3. Lastly, the *rentier* may decide to hold out, with the sole expectation of increasing the sum the firm is willing to pay him for the property.[23] The "hold out" outcome alters slightly the game, turning

---

23. Clearly, here "holding out" is different from declining engagement: the *rentier* shows dissatisfaction with the firm's offer with the expectation that a second, higher offer will be forthcoming.

it into a contest of seduction—hence, the boudoir allusion—and attrition: if indeed the *rentier* is correct that the firm will pay a higher price in every round of bargaining, then his potential reward is also higher and, in turn, the firm's transaction costs increase accordingly.

Clearly, this game cannot be played too many times in the *rentier*'s favor, because at a certain round, the utility the firm extracts from the purchased site at an inflated price will be inferior to the reward of seeking an alternative site outside the quartier, or a less mercenary *rentier*. Let us not forget that this is a quartier of profit-maximizing egoists!

Axelrod discussed the best strategies that egoists can employ to resist invasion. If we continue, however, with the same scenario of entrepreneurial *rentiers*, we should also note what the best strategy may be for the firm to keep purchase-price inflation under control: looking again at the Boudoir Game, we note that in each round in which the firm has the initiative, it has the capability of defecting by choosing an alternative site outside the quartier, thus imposing a transaction cost of W on both parties. Emotions, in this case, work to support the consequentialist, utilitarian paradigm: employing the logic of backward induction, the *rentier* will make the determination that since the firm will defect once he has held out one too many times, it may be better to sell in the first opportunity for a reward of 3, than risk sucker's payoff rewards of 2, 1, or 0 represented by the "Avoid engagement" and "Alternative site" outcomes.

All this, however, discounts the importance of community feeling. The two-person game between a real-estate corporation and a *rentier* who operates on the basis of a philosophy of community conservation will proceed as follows (fig. 37).

In the challenging business environment of a community united by a common agenda favoring landscape preservation, a firm (C) is again confronted with the choice to penetrate a less hostile community, thus, securing a lesser reward, which is further diminished by the act of surrender. In this case, the resident homeowner (R)—or best, the community as a whole, since the ultimate goal is the protection of a public good—enjoys a high reward of 4, representing the untampered-with state of the quartier's landscape. Alternatively, the firm can seek market engagement with the resident homeowner. The resident has three options: she can sell the property, in which case the firm extracts a high reward of 4, while the community suffers defeat by allowing the firm to gain a territorial advantage, and by having one of its members defect. As we will see later in the study, single defections can be important to the integrity of street blocks,

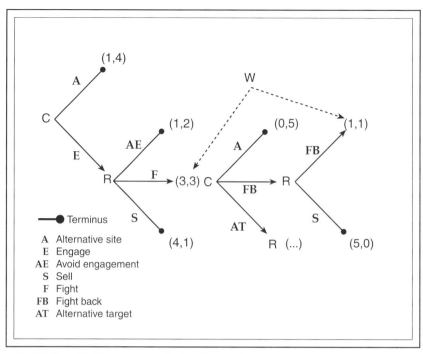

Fig. 37. Guerrilla Game: community-based reasoning

since once developers appear to take over a critical mass of the block, then the remaining properties usually are converted or demolished rapidly. The resident can, alternatively, choose passive resistance and not engage in any kind of negotiation with the firm, which will result in a reduced reward of 2 for the community, and a continuing dead end for the firm. Lastly, she can fight the firm alone, or with the assistance of members of the community, just as the Ateliers Mommens have attempted. This outcome represents greater risks and enhanced opportunities for both players: on the one hand, it gives the developer entry into a process which may yield victory (purchase, forced sale, demolition, etc.). On the other hand, it gives the resident homeowner/community the opportunity to defeat the firm in court or in the public opinion forum and create a basis for further activism. I therefore assigned an equal reward of (3,3) to both, since the "Fight" outcome leads conditionally to the highest reward in the second round of the game.

In the second round, the firm can choose to retreat from the quartier, thus receiving the sucker's payoff of zero, while rewarding the community with the highest reward of 5. Alternatively, it may select another homeowner as target in the hope of discovering a defector. This move will lead to the first round of a new Guerrilla Game. Lastly, the firm may decide to "Fight back," by employing its extracommunity resources in government, financial institutions, and the planning establishment. The second round is then completed with the response of the homeowner, who will succumb to pressure, sell the property, and thus receive what the community considers the sucker's payoff of 0. The developer receives the terminal ultimate reward of 5 for taking control of the property and defeating the community in the name of the homeowner. Alternatively, the community can continue fighting back. The reward pair of (1,1) suggests that the actors extract less and less utility with every round of hostilities, as, on the one hand, the community rallies against the developer to the detriment of future mutual cooperation, and on the other hand, both accrue transaction costs.

It is critical here to see how an ideological filter alters the reading of the motivations. If we adopt community-based reasoning, the sale of the property in the second round represents the sucker's payoff. If, however, we employ instrumental logic, then the sale of the property—which yields a monetary reward to the owner in an environment of scarce supply of land—is seen as both consistent with instrumental self-interest and desirable. In other words, one can easily then interpret the sale as the act of a free rider: a defector who succumbs to the temptation to sell when everybody else resists. Prices, and therefore her reward, will be the highest, since the number of properties in the market will be well below corporate demand. The action would constitute free riding, since the defector enjoys the bounty of a pristine neighborhood landscape, while cashing in on the sale. There is, of course, a finite number of defections possible before the rewards of free riding are eroded: too many defections increase the supply of available properties and therefore their market prices; and too great a number on the portfolio of real-estate developers means that the landscape is under significant modification.

The critical question, then, is whether the quartier Européen-Léopold fits best in the Boudoir or the Guerrilla game. There is good evidence that the quartier residents have functioned more frequently within the Boudoir Game, while initiatives for conservation and regimes of resistance have been the exception. In fact, the mass movement for landscape preservation in Brussels represented by groups such as ARAU does not have its roots in the quartier Européen-Léopold, although several of its current battles are being fought there. Ultimately, what I have attempted to do by introducing game-theoretic reasoning and the consideration of spatial linkages with international economic and political structures to the discussion of urban projects, is to build on Logan and Molotch's model of place entrepreneurs and explore the tension between agency and structure in the formation of a new urban form. Below follow what I believe are critical data on the character of these structures.

The State/Private-Sector Cooperative Regime

Thus far, we have presented the impact of the constitutional reform on Brussels's institutions and power structure, the nature of national and Brussels Regional governments' policies as they relate to keeping the European Communities in Brussels, and the ways in which these policies are driven by the uncertainties involved with the status of Brussels as the capital of the European Community. One more participant is necessary to complete the power tableau: private business interests, and in particular a small number of corporations with strong links to the various levels of Belgian and Brussels government.

The participation of the private sector in building the necessary infrastructure is important for financial and legal reasons: financial resources are relatively scarce, especially during a worldwide economic recession. As a result, the Belgian state is faced with a political dilemma. On one hand, if it does not expend the resources to provide the conditions for attracting the European institutions to Brussels, the European Communities would be dissatisfied with the working and living environment and would be less compelled to stay. The departure of the Communities from Brussels would have significant political and economic repercussions. On the other hand, if financial resources and political influence are lavished on the urban elements of a European strategy for Brussels, other politically sensitive areas of policy in the Brussels-Capital Region are threatened with neglect. Housing price increases that are outstripping cost-of-living raises, and traffic and pollution from the growing number of automobiles in the capital are the two most sensitive issues. The political repercussions resulting from their neglect may prove as hazardous for the ruling partisan coalition as would those resulting from the departure of EC headquarters.

The legal dimension is equally important to the process. Since Brussels is not the de jure seat of any of the European institutions, the

Belgian state is barred by Community law from building any installations for the European institutions "temporarily" situated in
Brussels.[24] The rights of the Belgian state are narrowly defined and
allow it to act only as landlord, or lessor of buildings and land, but
not as initiator or financial sponsor of projects explicitly designed for
use by the European Communities.

The venue for a state/private-sector cooperative regime on the issue of "building Europe in Brussels" is clear: the private sector
undertakes those projects that the Belgian state in its various guises
cannot.[25] Since the building of Europe in Brussels has become an immensely profitable enterprise, the private sector is very happy to
oblige the Belgian state, and at the same time nourish the important
political relationships that make things happen in Belgium (fig. 38).

Cooperative regimes resembling urban coalitions between
business and local authorities are very common in the running of
cities in market economies. Legitimate means of self-promotion
include urban coalitions of politicians and various kinds of businesses, and the resulting boosterism, which helps a city compete
against other cities for capital, skilled labor, tourists, the conference
trade, and amenities. Brussels is in competition with at least two
other cities: Luxembourg and Strasbourg. Strasbourg is responding to
the challenges mounted by Brussels by building its own Parliament
installations. Moreover, with the unification of Germany, Bonn
appears an attractive site with an abundance of modern green-field
office developments. The French and other Europeans may find
Berlin an even more unpalatable choice than Brussels, but the
economic muscle of Germany and the opening of economic markets
to the East may tip the balance in its favor. In light of such tough
competition, boosterism of the kind fostered by the Belgian
authorities and corporations should be viewed as legitimate.

---

24. Council of Ministers' agreement in the Maastricht summit of 1981.

25. There is nothing strictly illegal about this process. The Belgian state observes the
wording of the 1981 Maastricht agreement to the fullest, while successfully pursuing its
European agenda for Brussels. This "thesis" has been explored extensively by the liberal
press and the Greens of the European Parliament. The public feeling about this political
maneuvering is very tame. As Paul Staes, maverick Belgian member of the European
Parliament for Agalev (Alternative Living), the Flemish Green party, admitted, he is
the only Belgian MEP who openly opposes the candidacy of Brussels as permanent seat
of the European Communities. In January 1989, he was the only one of the twenty-four
Belgian members of the European Parliament to vote against Brussels as the seat of the
European Parliament. Staes, p. 34.

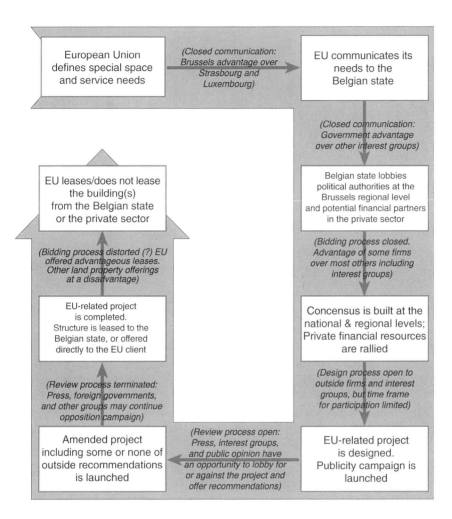

Fig. 38. The life cycle of a European construction project in Brussels

If we look at a possible cooperative regime between Belgian state authorities and private firms in terms of a rational-choice game, or a Prisoner's Dilemma, we arrive at four scenarios (fig. 39):

1. Both parties agree to cooperate on the project (cell no. 1). The parties negotiate the division of labor and responsibilities, and account for the costs and benefits of the operation. By agreeing to cooperate, they reduce transaction costs involved in a

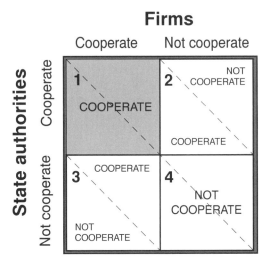

Fig. 39. Four game-theoretical choices of action for state authorities and private firms

lengthy search for alternative venues or partners, and mini-mize their financial and political exposure.

2. Both parties defect, or refuse to cooperate on the project (cell no. 4). In this case, parties are either structurally or politically incompatible, or in some way unable or unwilling to work to-gether. One of the partners may deem it more profitable to mount a competing project to the one proposed. Transaction benefits have been forfeited, and each defector has sustained at least some penalty.

3. In the case where one party seeks cooperation and another de-clines (cell nos. 2 and 3), cooperation is obviously also impossi-ble. According to rational-choice thinking, the defecting party deems it profitable to remain outside the proposed regime and either undertake the project alone or perhaps seek other part-nerships. In the textbook case, the party defects because the reward associated with the defection appears to outweigh the benefit of cooperation. Transaction costs and risk are highest, of course, for the party which was poised to cooperate. For every additional partner or set of partners, the number of scenarios increases rapidly, while there remains only one opti-mum cooperation scenario (fig. 40).[26]

---

26. The traditional Prisoner's Dilemma involves pairwise interaction, although a sin-

In the case of Brussels and the quartier Européen-Léopold, the primary players (government and certain players in the private sector) seek efficiency and advantage by limiting the access of others to the planning process. These other potential participants are independent planning groups, such as ARAU or BRAL (Brussels Urban Environment Council), resident committees from the quartiers Léopold and Nord-Est, and a variety of Belgian and foreign real-estate development and construction firms. Alternative planning agencies may be kept out of the process because they are found to have fundamentally different planning priorities and ideologies. There may be a certain amount of legitimate expediency to this course of action. If the evidence suggests however that most *firms* are excluded through the structuring of the publicity and bidding processes, then we can assert that the EC urban projects are the result of inefficient and noncompetitive practices. Whether these practices are in fact unethical or illegal adds to the complexity of this urban maze.

Conspiracy Uncovered, or Business as Usual?

Brussels has been the scene of grand urban changes since Belgian independence in 1830, and the nature of these changes has been influenced by the role of private capital. The contemporary urban morphological frame of Brussels features public and private projects that break sharply with the original nineteenth-century one. These projects have historically served the interests of those building alliances between government and the private sector. The end result is a seemingly incoherent urban morphological frame mixing building types and architectural styles with little consideration to the aesthetics or functionality of the affected neighborhoods. This phenomenon is so characteristic of Brussels that when a city exhibits this kind of erratic morphology one could claim that it is undergoing "Bruxellisation" (fig. 41).

---

gle player may be interacting with a greater number of actors as suggested here. Olson, Harding, Schelling, and Taylor have modeled the more complex *n*-person Prisoner's Dilemma in which the principal goal is the optimized disbursement of collective goods. Mancur Olson, Jr. *The Logic of Collective Action* (Cambridge, MA: Harvard University Press, 1965); Russel Hardin, "Collective Action as an Agreeable n-Prisoner's Dilemma," *Behavioral Science* 16 (1971): 472–81; Thomas C. Schelling, "Hockey Helmets, Conceled Weapons, and Daylight Savings: A Study of Binary Choices with Externalities," *Journal of Conflict Resolution* 17 (1973): 381–428; Michael Taylor, *Anarchy and Cooperation* (New York: Wiley, 1976).

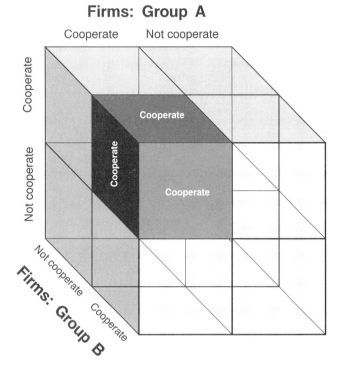

Adding a partner or set of partners, doubles the number of scenarios. Only one in eight possible combinations signifies the formation of a regime of cooperation.

Fig. 40. Cubiform Game

The nineteenth-century projects, which removed much of the medieval urban fabric, aimed to sanitize and modernize the industrializing city (for example, the diversion and canalization of the Senne river, the extension and widening of boulevards through dense neighborhoods of townhouses with street-level shops, the building of the north and south train stations, the residential expansions of the city into the countryside, and the engulfing of surrounding municipalities). Industry found its home in the western lowlands of the city in proximity to the canals and the port. Financial surpluses were put to use in the founding of residential projects such as the quartiers Léopold and Nord-Est, and in the opening of the avenue Louise.

Fig. 41. The impact of Bruxellisation on the Pentagon. View of the southern part of the Pentagon as seen from the place de Brouckère. From left to right, the gothic spire of City Hall at Grand Place, a twenty-four story office at the place E. Vandervelde (middle ground), the dome of the beaux-arts Bourse (foreground), and its contemporaneous Palais de Justice.

The twentieth-century construction of the central station in the heart of the Pentagon city along the underground rail line connecting the north and south stations and the construction of the massive Cité Administrative over this rail yard constitute examples of heavy-handed urban planning, where government and private interests have come before the interest of the residents, urban aesthetics, and the preservation of the city's architectural heritage (fig. 42).[27] The criticism over the Dallas-like coldness and austerity of the Cité Administrative which replaced eighteenth- and nineteenth-century neighborhoods is not the exclusive domain of the liberal circles. The vast plazas and ramps of the Cité Administrative remain underused and underpopulated even by the thousands of government officials who used the office buildings. This purely office district has been used by the press to warn developers of what the quartier Européen-Léopold may look like in the near future. With industry leaving the city at an accelerated pace since the 1960s, the tertiary sector has become the driving force behind large-scale urban change. The projects where finance capital is dedicated are usually not residential—though these exist as well—but mostly service- and office-related. The rates of growth and rates of return for office properties are significantly higher than those for residential or even commercial properties. Given the traditional cooperation regime between government and business, the contemporary urban planning tendency is very much in favor of the office sector.

According to Georges Timmerman, independent journalist and regular contributor to Dutch-language newspapers and magazines, certain large construction companies, speculators, and real-estate developers based in Brussels function in symbiosis with local government authorities (Regional, metropolitan, and municipal) and the Social-Christian Party (PSC/CVP). The glue of this alliance is purported to be the financial support of the party by these corporations: The De Pauw and Blaton Groups, Etudes et Investissements Immobiliers (EII), Compagnie Immobilière de Belgique (Immobel), Société Belge des Bétons (SBBM), and Bernheim-Comofi.[28] Financial

---

27. The removal of the traditional city for this project has been deemed a poor choice by urban historians. Thierry Demey, *Bruxelles: Chronique d'une capital en chantier; Du voûtement de la Senne à la jonction Nord-Midi* (Brussels: Paul Legrain/Edition CFC, 1990), pp. 235–40.

28. Timmerman, pp. 8–9. The Social-Christian Party represents the conservative political wing. Timmerman asserts that it is mainly the Brussels branch of that party, and especially, its Walloon leadership (PSC) which facilitates the EC projects.

Fig. 42. Nineteenth- and twentieth-century incarnations of the nationalist city

backers of the real-estate projects include the Générale de Banque, Banque Bruxelles Lambert, Kredietbank, Paribas, the pension funds An-Hyp, BAC, CERA, the Prévoyance Sociale, prestigious smaller banks such as the Caisse Privée, and Banque Degroof (bankers to the royal family and the nobility), and finally an array of Belgian insurance corporations, such as the groupe AG, Royale Belge, Assurances Populaires, and JOSI.[29] Conspicuously absent are large British real-estate development and management firms, such as Jones Lang Wootton, Richard Ellis, and Healey & Baker. These have a significant presence in the quartier Européen-Léopold as owners, lessors, and managers of large office properties, but appear to have kept out, or have been kept out, of the actual construction of the EC projects of the Centre International du Congrés, and the new Council of Ministers complex on the rue de la Loi (fig. 43).

The group of corporations belonging to a variety of sectors, and overwhelmingly representative of Franco-Belgian financial interests, should be considered as subscribing to a single urban development ideology in the making of the quartier Européen-Léopold. On the one hand, firms are, not surprisingly, expected to gravitate to high value-added projects. Particular political, family, and especially ownership relations are additionally responsible for knitting a significant number of these players into an informal and like-minded consortium. The manner in which the involvement of these firms in the urban development process has been described by the liberal press suggests that perhaps the dramatis personae discussed below are engaged in unethical if not downright illegal practices.

Georges Timmerman notes that among the private-sector players in the quartier Européen-Léopold, the company Etudes et Investissements Immobiliers (EII) stands out as the premier supplier of rented office space to the European Communities. EII finds itself in this enviable position because of historical circumstance. When Paul Henri Spaak, Belgian foreign minister and champion of the idea of a united Europe, was researching locations for the siting of the European Community institutions in the early 1960s, Armand Blaton, at that time chairman of the board of EII, suggested to him a block of fin-de-siècle houses between the streets Loi, Taciturne, Joseph II, and Charlemagne avenue in Brussels. At Spaak's recommendation, the Belgian state adopted the idea and erected the Charlemagne Building, which has since housed the Council of Ministers.[30] This first in a long series

---

29. Ibid., p. 9.
30. Ibid., pp. 41–42.

Fig. 43. The hemicycle of the Centre International du Congrés under construction (September 1991), as seen from the rue de Trèves. Construction of the accompanying office towers continues today.

of buildings used by the Communities constituted the initial magnet for the urban transformations in the quartier which followed.

In 1963, the A. Blaton et Fils company and the Compagnie Foncière Internationale (CFI) were equal partners in EII. For the next two decades, the former was the principal contractor for EC projects, while the latter was the principal financier. As EII proved immensely successful, it was taken public by its two parent firms in 1987 and quickly became a treasured stock in the Brussels exchange in the heat of the "1992" craze.[31] It is of note that while the participation of the two original owner companies appears to have decreased with stock sales since 1987, the principal buyers of stock are companies that are closely linked to them. For example, the current controlling shareholder is the Société Foncière Internationale (SFI) (40 percent), which is in turn controlled by the Paris-based Compagnie Foncière Internationale (CFI), itself a subsidiary of the Compagnie Financière de Suez. Other shareholders include La Hénin Nord, another subsidiary of Suez, the Compagnie Immobilière de Belgique (Immobel), and the Société Générale de Banque, a financial player in Brussels real estate since the nineteenth-century. The loop is completed with the Société Générale being a subsidiary of Suez (fig. 44).

The Count Bruno Dadvisard, chief executive of SFI and EII, makes the case for investing so heavily in Brussels real estate and especially in the quartier Européen-Léopold:

> An underlying strength of our operation is our collaboration with professionals of the highest standing . . . We participate in the real estate market with the support of partners who are well positioned in the market to know about local prices, legislation, and long-term possibilities. In any case, [we work with] persons who are very knowledgeable about all the important factors which influence real estate. In Belgium, EII works with the *groupe Blaton*, to which I would like to relate my respect for twenty-five years of collaboration.[32]

Dadvisard continues by noting the strong linkage between their interest in Brussels real estate and the growth of the European Communities there.

---

31. *Compagnie Foncière Internationale*, brochure about the public offering of EII stock (Brussels: CFI, 1990).

32. Speech of Bruno Dadvisard at the Brussels Stock Exchange on the occasion of a public offering of stock.

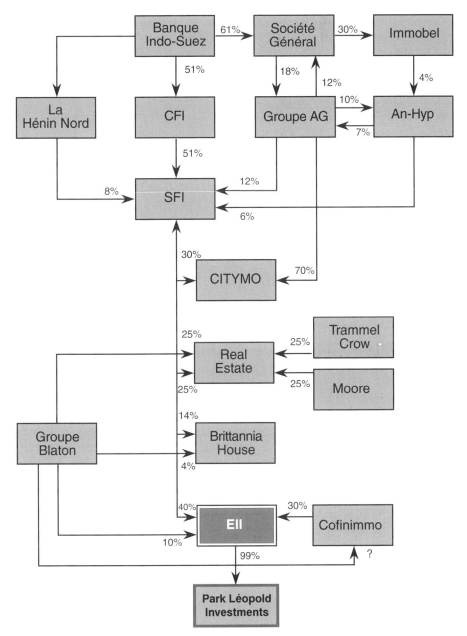

Fig. 44. Corporate linkages of Etudes et Investissements Immobiliers (EII). *Source:* Georges Timmerman, *Main basse sur Bruxelles: Argent, Pouvoir, et bêton* (Brussels: Editions EPO, 1991), p. 51.

We have established a strong presence in Brussels where the concentration of tertiary sector services has created a great demand for office space . . . It is because of the insufficient extent of zones reserved for the office sector that the tertiary sector is leaving the areas traditionally reserved for offices. The number of [office development] projects currently under way are increasing. [Still,] offices are scarce in Brussels and that is one of the principal reasons why the prices are rising. This phenomenon is further reinforced by the fact that the institutions of the European Communities will, without doubt, be situated permanently in Brussels, and with them a great number of [related] businesses.[33]

The link between the SFI and EII investment and the growth of the European institutions in Brussels is clear. The area for which tertiary sector businesses are leaving traditional office zones is, clearly, the quartier Européen-Léopold. Dadvisard's claim about office scarcity is, of course, subject to criticism.[34] The nature of SFI and EII holdings leaves no space for doubt: shopping malls 20 percent, office buildings 72 percent, among them the Charlemagne Building, the buildings of the political groups and committees of the European parliament on the rue Beliard, and the two massive projects under construction, the Centre International du Congrés (CIC) and the new complex which will house the Council of Ministers. Park Léopold Investment, a subsidiary of EII, appears to have been created to build the CIC.[35]

The particular individuals managing these companies can often be readily identified. With the exception of the Banque Indo-Suez and the Société Générale de Banque, which are world-class conglomerates, most of these specialized companies—many of them privately held—are of small or medium size by Fortune 500 standards, however cash-rich or influential they may be in the Brussels market. The name Blaton figures prominently in this market. One of the Blaton companies—Bâtiments et Ponts et Cie (BPC)—participated in the consortium that built the Charlemagne and Berlaymont buildings of the EC. French financial interests have been important at Blaton: 75 percent of BPC has been purchased by the French Compagnie

---

33. Ibid.

34. ARAU and IEB are on the record disputing it.

35. Timmerman, p. 44.

Générale des Eaux (CGEaux) through the intermediary of its subsidiaries, the Compagnie Générale de Bâtiments et de Construction (CBC) and the Compagnie Générale Européenne (CGE) (fig. 45).

Both the EII and the Blaton networks of companies represent the interests of powerful individuals located in Brussels and Paris. These persons have links with the Palace and with banking circles, and nourish relationships with relevant metropolitan, Regional, and national-level politicians. Much as Georges Timmerman wants to expose these patronal networks of firms and persons as illegal, they bear some resemblance to city machines that have emerged from time to time in U.S. cities.

The quartier Européen-Léopold regimes, however, are not Belgian incarnations of urban coalitions or city machines, as we understand these concepts in the United States. For one, the quartier Européen-Léopold covers only a fraction of the Brussels metropolitan area in spite of its massive significance for the development of the city as a whole. Second, the regimes do not involve local elites exclusively. In fact, the catalyst/client—the EC—is not native at all. Finally, the objective is not the perpetuation of a partisan regime and the device is not electoral patronage, although patronage of sorts is at work in the quartier regimes as well.

Therefore, they should be viewed as a new device of urban boosterism, reflective of global financial processes and the European enterprise of integration. There is often not much illegal or even unethical about these regimes, especially when the city is made to work well. The possibility for manipulation of city affairs and urban planning by individual interests is, however, real. The *process* that generates the urban regime we have just described may, hence, be less a problem than is the profit-maximizing agenda of its participants.

The fight over the shape of Brussels in general and the quartier Européen-Léopold in particular centers on the tension between the desire for a quality urban environment and the maximization of economic benefit. This polarity is common to many cities in Western Europe and the United States. Having put the smokestack city behind them, urban leaders attempt to capture high value-added tertiary and quaternary industries. In the Brussels context, this particular polarity is underlined by suburbanization and the divestment of private investors from the secondary sector. The long-standing contention of Regional government and the inhabitants of the city is that housing and urban conviviality should come first; the response of the

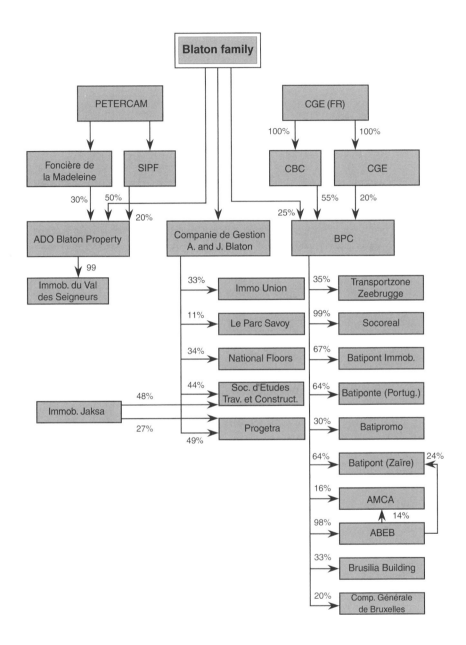

Fig. 45. Corporate linkages of the Group Blaton. *Source:* Georges Timmerman, *Main basse sur Bruxelles: Argent, Pouvoir, et bêton* (Brussels: Editions EPO, 1991), p. 52.

private sector is that constraints of the private sector's urban activities would simply add business flight to population flight—and where would Brussels be then?[36]

The Brussels urban coalition or machine brings together firms, political institutions, governmental organizations, and their client planning groups. The framework is institutional, but it is often the political or economic profit of individuals that directs the institutional framework. As Paul Staes, member of the European Parliament describes, the building permit for the Centre International du Congrés was "a surrealistic piece of political/administrative maneuvering: . . . Jean-Louis Thys, secretary of state for the Brussels-Capital Region, applied to himself for a permit and actually managed to get it" (see the Appendix for the complete text of the building permit). Although the courts reversed the case, the end result was that the CIC was approved through a compromise between the real-estate promoters and residents' committees.[37] Next to Thys, Staes and Timmerman single out Paul Van den Boeynants, head of the Association for the International Promotion of Brussels (APIB) and the Brussels International Trade Fair nonprofit association, and former minister of defense. His close ties to Immobel, the Société Générale, the Etudes et Investissements Immobiliers, and the moneyed circles of Brussels have been documented. The associations he heads have been described in the liberal press as fronts for his interests.

If this is indeed the truth, then we can try to answer our original theoretical question about structure and agency. The evidence is, regrettably, not conclusive, but suggestive. First, investigating a single and perhaps idiosyncratic case like Brussels does not allow us to draw extensive generalizations. The maneuverings of firms and politicians do certainly occur in other European and American cities, but would clearly serve very different goals from those special EC ones of Brussels. If in spite of this caveat we wanted to place the political geography of Brussels somewhere on the "structure–agency" continuum, we would be right to place it slightly off center in favor of structure. Structures, such as firms and political institutions, serve as indispensable instruments in the implementation of Brussels's urban

---

36. P.-A. De Smedt, "Bruxelles entre son avenir international et celui de ses habitants: Exposé du Président de l'Union des Entreprises de Bruxelles," *Forum Immobilier* (15 and 16 March 1990).

37. Staes, p. 17. Another example of such maneuvering involves the architectural plans for the CIC, which contained the date of the urban planning certificate—which was not issued until later (27 April 1987).

policy, in spite of the fact that individual agency, as exemplified by key persons like Thys and Van den Boeynants, is essential to the running of the structures, which are largely political operations. Structures and persons carry out a policy agenda by building regimes of cooperation which last as long as common interest exists.

These regimes perform a number of spatial tasks: thanks to their ability to rally capital and political resources, they invest particular pieces of territory with locational significance, thus turning them into special *axes mundi* around which the quartier land market revolves. The new Council of Ministers complex and the Centre International du Congrés, and historically the Berlaymont, constitute such *axes*. Regimes can be described as "software" to the built environment's "hardware." They provide the routing scheme for the performance of a specific task that invariably has a material impact on the quartier. Most emphatically, they are more than mere blueprints, because their very essence lies in the abstract transactions between agents rather than in plans on paper. Regimes also represent gateways—or, perhaps more appropriately, revolving doors—through which economic and political agents and structures originating at the local, Regional, national, or transnational levels participate in spatial decisions: they contribute capital, add to or alter the functional character of the quartier, and bring sometimes alien planning, technological, architectural, or aesthetic sensibilities to the existing growth culture of the quartier. Finally, the quartier regimes are mirrors of comparative advantage and economic opportunity in, again, a variety of spatial scales: they encapsulate "the possible," or better, "the feasible," in profitable urban development at the lot, block, quartier, and city levels. Each regime constitutes an ephemeral forum for self-interested agents, in which land-use development and management are mediated through market brokerage and political patronage.

The Newly Empowered "Weak," and a Look to the Future

Residents' committees and public interest groups seek to moderate the excesses of the urban machine by advertising their discontent (fig. 46). They often win victories, although generally small ones: stopping the demolition of a group of neoclassical houses, or getting a threatened façade into the highly limited national registry for protected buildings. They are losing the big battles: limiting the expansion of the office sector in the quartiers Léopold and Nord-Est, promoting mixicity of functions in the city, and keeping housing prices

Fig. 46. "Sauvez-moi, reagisser." *Top:* Graffiti against the demolition of the fin-de-siècle rowhouses at the corner of the boulevard Charlemagne and the square Ambiorix. *Bottom:* Graffito against the advance of the EC office park at the place Jourdan.

under control. They are clearly building their own regime of cooperation—a regime of resistance—to take on the machine. Only if these groups are able to take control of urban planning, however, are the current development trends likely to be arrested. This is highly unlikely to occur.

8

In Search of Monumentality
in Consumption

T HE EDITORS of the *Architectural Review* admonished the builders of the Centre International du Congrés in an April 1993 piece titled "Outrage." They ask: "Is this the way to make the Parliament of the European Community? Surely the supreme legislative body of a continent deserves better than a glazed behemoth."[1] Considering that the editors are unaware of the byzantine urban strategies of the quartier, their accusation is extremely befitting: Not only is the CIC of questionable architectural and aesthetic value, it also fails to capture in built form the historical significance—though not the economic expediency—of the European integration enterprise. Surely building volume alone cannot capture magnificence of this sort. Surely this is a case where the architects of the buildings have failed the architects of the institutions.

I make two general assumptions: one, that today Brussels should be viewed as the de facto capital of the European Community and judged as such, even if the relevant treaty provisions consider its status as capital merely provisional. Two, that capitals have historically been conceived and built with more than the machinery of state in mind. Cast as monuments to the persons, ideology, and political order of the nations that have given rise to them, they are mirrors of culture and society and have been forged to endure. In this and any discussion of aesthetics, the taste of the discussant is inevitably revealed and ultimately pits him against those who do not share it. This discussion is unlikely to please anyone shaping the quartier Européen-Léopold today.

The complexity of the quartier Européen-Léopold's pedigree allows each of the parties involved in its shaping to accuse the others of

---

1. "Brussels Behemoth," *Architectural Review* 193, no. 1154 (April 1993): 15.

failing to produce a first-rate product. The limited planning put forth
by the Belgian authorities has come under assault from both profit-
maximizing business circles and conservation-minded civic groups:
The former criticize it for distorting and inhibiting the operation of
market forces that allegedly allocate space to users in the most efficient manner. The latter denounce it as the tool of the urban developers and their allies in government, and as a front allowing for the
piecemeal dismantling of the remaining nineteenth-century city.
Market forces and profiteering by different groups of place entrepreneurs have been blamed by the liberal press, by longtime residents of the quartiers Léopold and Nord-Est, and by their champions,
ARAU and IEB, for creating a monofunctional, Dallas-style business
district in place of a diversified and aesthetically and historically
valuable Brussels neighborhood. The response of the urban developers is that they work within an inefficient system lacking in leadership and vision and wrought with cumbersome bureaucracy and
regulations. If Brussels will not allow them to build an appropriately
functional quartier for the emerging European union, then perhaps
Paris or Bonn, with plentiful empty office space, may oblige them.
The result is an urban form which is receiving good scores for its performance in a narrow band of administrative and business functions
within the city of Brussels, but scores poorly as an integrated part of
an urban whole and scores abysmally as a showcase of European urban ingenuity.

Thus everyone appears to agree that the end product, the quartier
Européen-Léopold with its canyon-like rue de la Loi and quaint rue
Pascale, does not evoke the splendor and historical significance of the
dawning united Europe. The quartier fails to encapsulate in its appearance the monumentality of the task the bureaucracy and its spin-
off industries ably perform in the unremarkable office towers they
occupy. The quartier Européen-Léopold comes indeed as a sad conclusion to an intermittent but illustrious tradition of European capitals
begun with Julian Rome, recaptured in tenth-century Aachen,
apotheosized in Haussmannian Paris, and interrupted by the otherwise fortunate bombing of Berlin in 1945. Although the quartier
Européen-Léopold should not be required to evoke any or all of these
European urbanistic traditions, it may be a reasonable assumption
that the likely capital of Europe should do one of two things: either
glance back and reflect the richness of European urbanism and the
material tradition that together evoke a sense of common origins
among Europeans, or look ahead to a presumably brighter future of
democracy, cultural unity, urban innovation, and artistic genius and

break with tradition. The Brussels quartier of quick and dirty façadisme and profit-driven megaprojects does neither and proves disappointing as the symbolic pivot of Europe.

Urban Morphology

From the Baroque Grand Place to the Modernist Quartier Européen-Léopold

A promenade from the heart of the medieval city inside the Pentagon to the quartier Européen-Léopold can be used as an effective shorthand for the evolution of Brussels's urban morphological frame. The visitor will start in the Grand Place, which was the focal point of city politics until nationalist independence in 1830, and will walk through areas transformed during the nineteenth and early twentieth centuries to reach the limits of the Pentagon city at the intersection of the avenue des Arts and the rue de la Loi. The visitor will then traverse the boulevard du Régent and the avenue des Arts, which separate the Pentagon from the first nineteenth-century extension of the city, and enter the quartier Européen-Léopold. The rest of the tour will offer the opportunity to compare form and existence in the two cities bordering on the avenue des Arts.

Visitors should start their tour at the Grand Place, the magnificent and magnificently preserved late-seventeenth-century square in the center of the medieval Pentagon city. Reconstructed by the citizens of Brussels following the bombardment of the city by the army of Louis XIV in 1695, the square exhibits an architectural style based largely on the Italian baroque. Exceptions include the City Hall and the so-called Maison du Roi, which were constructed in the gothic style. The architects of the houses surrounding the square took classical architectural elements one step further to reflect the intensity and luxuriance of Flemish baroque. Columns, capitals, and pediments in the Doric, Ionian, and Corinthian orders and their Roman derivatives contribute to lavish decorative façades that include narrative sculptural bas-reliefs, busts, statues, medallions, vases, torchères, scrolled frames, cornucopias spilling fruits and flowers, and gilded inscriptions.[2] Most houses have characteristic Flemish and Dutch gabled or stepped roofs in contradiction, again, to the Italian baroque archetypes. The façade ornamentation serves as a frame around severe lead-latticed rectangular windows.

---

2. Author's fieldwork with the assistance of Des Marez, p. 42.

The Grand Place of the 1990s is the focal point of the transient tourist trade and the people and businesses who cater to it. Cafés, restaurants, lace and chocolate shops, a major bank, a travel agency, the city's tourist information agency and, perhaps surprisingly, a pet-supply store line the rez-de-chaussées of the square. The cobblestone streets host a weekly flower market and an occasional bird market. Service trucks deliver to the businesses in the early morning hours and, regrettably, private automobiles are allowed most hours through the square.[3] Surrounding the square are the few sinuous streets dating to the city's medieval past. One of them, the rue de la Colline, leads from the northeast corner of the Grand Place toward the carrefour de l'Europe.

The first mention of the rue de la Colline dates to the thirteenth century. Establishing the link between the heart of the city and the *chaussée* leading to the high city and the Palace of Coudenberg,[4] the rue de la Colline survived over seven hundred years a major city fire, the 1695 bombardment of the center, independence, and the modernization of the Pentagon during the last two centuries. Lined with rowhouses from the seventeenth and early eighteenth centuries, the interiors of most of which have since been modernized to a vary-ing extent , the livelihood of the rue de la Colline is also defined by the tourist trade.

The rue de la Colline ends at the carrefour de l'Europe, the inter-section of the rues de la Montagne, du Cardinal Mercier, and de la Madelaine, and the base of the modest escarpment which divides the Pentagon into a "low" and a "high" city. The carrefour includes four street blocks and was an element of the construction project connect-ing the Midi to the Nord train station via an underground link (1935–52).[5] It occupies part of the old working-class quartier de la Putterie, which was demolished after 1910, and until recently undeveloped.[6] Looking at the course taken by the underground junction, it is ques-tionable whether the destruction of the Putterie was necessary or simply employed as a means to displace the poor from the modern

---

3. At press time, the city was trying a new, restrictive automobile traffic regime, which barred cars from the square and its environs.

4. Ministerie van de Vlaamse Gemeenschap, Bestuur voor monumenten en land-scappen; Ministère de la Communauté Française, Administration du Patrimoine Cul-turel, *Le patrimoine monumental de la Belgique: Bruxelles* (Brussels: Pierre Mardaga), 1: 286.

5. Ibid., p. 217.

6. Demey, pp. 34–35.

ized city center.[7] The Midi-Nord junction, the diversion of the waters of the Senne river, and the construction of the monumental building group of the Mont des Arts were instrumental in replacing the medieval ground plan with a planned, modern one.

After years of debate about its future, the carrefour is now partly occupied by three hotels owned by the French Ibis-Novotel Group. Citizens' groups led by ARAU have protested this transfer of ownership, which they claim has once more compromised affordable housing for the sake of corporate interests.[8] The ARAU has made a counterproposal which may provide housing in the carrefour through three different projects and may restore the building density which existed up to the turn of the century.[9]

All this would, of course, have been deemed entirely inadequate by proponents of coherent architectural aesthetics. Camillo Sitte, in his discussion of the restructuring of the square surrounding the neogothic Votivkirche in Vienna, points out that built surroundings of a different style detract from the visual coherence of the whole. He writes, "[I]nstead of mutually benefiting from the careful juxtapositions of their attributes and style, the edifices [in the Votivkirche square] each play a different melody . . . One has to have a particularly strong constitution to endure it."[10]

Opinions about the value of such rigid visual coherence and aesthetic standards have clearly changed over the last ninety years. This is fortunate, considering that the carrefour de l'Europe contains buildings—some of questionable aesthetic value—spanning the last four centuries, and placed without regard to the impact each has on the

---

7. ARAU, "Terminer la reconstruction du carrefour de l'Europe en contrebas de la cathedral, de la gare Centrale et de l'Albertine," Press conference, Brussels, 2 May 1991.

8. E. R., "Grands projets immobiliers," *La Denière Heure* (6 May 1991).

9. Another issue is the management of two vacant street blocks in front of the main gates of the gothic Cathedral of Sts. Michel et Gudule a few hundred feet north of the carrefour. L. W., "Rebâtir le carrefour de l'Europe: L'ARAU soutient trois projets de reconstruction de logements sur le site, mais veut en améliorer l'architecture," *La Lanterne* (3 May 1991). By claiming that these blocks should be built to restore the original ambience of the quartier, architectural historians echo the commentary of Camillo Sitte on the spatial organization of squares in northern European cities. Sitte wrote that unlike buildings in the Classical style, which benefit from large, open fields of view, the full view of gothic cathedrals was meant to be revealed at close proximity. Hidden by densely constructed housing, the visitor is supposed to be overwhelmed when he reaches the base and looks upward toward spires which seem to disappear into infinity. Camillo Sitte, *L'art de batir les villes: L'urbanisme selon ses fondements artistiques* (Paris: Livre et Communication, 1990), pp. 69–72.

10. Sitte, p. 158.

rest. In all fairness, the hotel group has built in a style evoking the seventeenth- and eighteenth-century surroundings. Moreover, the project is a clear improvement over the desolate parking lot it replaced. It is likely that the patina of time will integrate the brick and limestone hotels to the remaining ancient fabric, creating a physical density resembling that of the preindustrial city.

The rue de la Montagne is lined with beautifully restored baroque and rococo townhouses and leads uphill to the boulevard de l'Imperatrice, which follows the path of the underground rail junction and dissects the modern Cité Administrative (1959–78, Gilson, Lambrichs, Ricquier, and Van Kuyck, architects), north of the quartier.[11] The visitors have traveled no more than two hundred meters and are already outside the Brussels of the Flemish or dynastic era. The midrise buildings lining the boulevard de l'Imperatrice are contemporaneous with the Cité Administrative and evoke characterless American urbanism rather than native architectural types. In the foreground lie two empty street blocks directly in front of the main portal of the Cathedral of Sts. Michel et Gudule (fig. 47). To the right of the cathedral lies the 1950s mid-rise building of the postal service built in the modern style, and to the left the classicizing National Bank building, constructed on the site of the Hospice des Enfants Trouvés in 1860, and expanded in 1905 and 1949.[12]

Crossing the boulevard de l'Imperatrice, the rue de Loxum leads uphill to an *étoile* intersection of the rues Loxum, Cardinal Mercier, Cantersteen, Ravenstein, and Montagne du Parc. There begins the large, curved rue des Colonies, which continues to the intersection of the rue Royale at the northwest corner of the parc de Bruxelles. Created in 1908–9 as an important new path through the ancient built fabric, it was an element in the reorganization of the quartier de la Putterie, the opening of the Nord-Midi junction and the building of the gare Central. Lined with imposing mid-rise buildings in an eclectic style heavily influenced by the classicism of the French beaux arts movement, the rue des Colonies was home to banks and apartment buildings for the well-to-do Bruxellois.[13] Like the late-nineteenth-century buildings lining the boulevard Anspach, the buildings of this street evoke Paris, although the scale is modest by comparison.

At the top of the rue des Colonies, the visitors have also reached the top of the escarpment separating the low and high cities. The rue

---

11. Des Marez, p. 320.
12. Ibid., p. 245.
13. Mardaga, pp. 291–96.

Fig. 47. The gothic and the postmodern coexist on the boulevard de l'Imperatrice.

de la Loi lies before them in all its automotive grandeur. Its northern flank inside the Pentagon is occupied by the neoclassical Palais de la Nation (the Belgian Parliament), originally built in 1779–83 (Guimard and Sandrié, architects) by the Austrians as the seat of the administration of the Brabant,[14] and eventually becoming one of the key centers of political power during the nationalist era. To the south of the rue de la Loi inside the Pentagon lies the formal parc de Bruxelles, redesigned in 1774 by the French architect Gilles-Barnabé Guimard and perhaps also the Austrian architect Joachim Zinner.[15] Across the southern flank of the park lies the Royal Palace erected during the reign of Léopold II. Ahead, across the quartier Européen-Léopold, at the end of the rue de la Loi and at a great distance, the visitors can see the triumphal arch of the parc du Cinquantenaire.

Walking along the northern side of the parc de Bruxelles on the rue de la Loi we reach the rue Ducale, which runs along the east side of the park. Adorned by neoclassical mansions, the rue Ducale was designed by Guimard in 1776. Most of the mansions between the rue de la Loi and the rue Zinner were originally commissioned by the Abbey of Deleghem and housed exiled French aristocrats after the fall of the ancien régime. In the early 1800s, the rue Ducale became a center for wealthy expatriate English, and until the twentieth century the mansions continued to be occupied by aristocrats. Today several of them house embassies and Belgian ministerial services.[16] The park and its neoclassical surroundings encapsulate the age of absolutism and the rise of nationalism in the city. One block to the east, the boulevard du Régent and the avenue des Arts extend to the site of the fourteenth-century fortifications. These define the limits of the Pentagon city and mark the beginning of what was the first nineteenth-century extension of the city: the quartier Léopold, now transfigured into the quartier Européen-Léopold (1850–1993).

From the corner of the avenue des Arts and the rue de la Loi, there stretches the canyon-like rue de la Loi, which ends at the western limit of the parc du Cinquantenaire. Almost the entire length of the rue de la Loi is lined with mid-rise office towers (five to seventeen floors) dating from the late 1950s to today.

The European Communities made their appearance assertively in the northeastern part of the old quartier Léopold with the construc-

14. De Marez, pp. 281–82.
15. Ibid, pp. 266–67.
16. Mardaga, p. 388.

tion of the Charlemagne Building (1965) at the intersection of Loi-Charlemagne, and the Berlaymont Building (1966) directly east of the first building on the Schuman roundabout.[17] Today, more than twenty-five years later, their presence is not as point-specific as it was in the 1960s, but is dominant throughout the quartier Européen-Léopold. Their influence has been pervasive in the evolution of circulation patterns, building types, and land use.

After two decades of spreading their facilities throughout the quartier, the Communities appear to be consolidating their operations in close proximity to the original point of entry: between the rue Stévin (N), avenue de Cortenberg and avenue d'Auderghem (E), rue d'Arlon and the place de Luxembourg (W), and the place Jourdan (S), around a renovated Berlaymont, the new Council of Ministers building, and the Centre International du Congrés.

Of greatest interest are the changes in morphology that have taken place in the quartier during the intervening twenty-seven years, and the bearing these changes have on the aesthetics of the city in general and the quartier in particular. Of critical importance is the link between the evolution of the market and the evolution of the built environment of the quartier. As we have noted in the previous chapters, the building culture of the quartier is largely driven by the private sector. Using data from a building census[18] and air photographs of the quartier obtained from the archives of the Institut Géographique National, I shall reconstruct the morphological changes which took place between 1957 (signing of the Treaties of Rome) and 1992.

In chapter 3, I discussed the morphological evolution of Brussels and the quartier in terms of three morphological periods: the dynastic, the nationalist, and the internationalist. We can divide the third morphological period into three time frames of special meaning to the aesthetics of the quartier: the launching of the Communities (1957–66), the consolidation of the office function (1967–85), and the relaunching of Europe (1986 to date). Each time frame describes a measure of morphological transformation of the quartier.

---

17. "Rapport Spécial de la Cour des Comptes relatif à la politique immobilière des institutions des Communautés Européennes," *Journal Officiel des Communautés Européennes* 221, no. 2 (3 September 1979): 11.

18. Author's fieldwork.

The Launching of the Communities (1957–66)

Air photographs taken on 7 April 1950 and 23 May 1956 illustrate the morphological stagnation of the quartier.[19] The differences between the two photographs are indeed minute (figs. 48–49).

In both photographs the ground plan established between 1850 and 1880, when the quartiers Léopold and Nord-Est were developed, remained intact. The quartier Léopold was designed in the shape of a rectangle with a northwest-southeast orientation. It was characterized by a rectilinear street pattern which would have allowed for equal access around the quartier. Adapting to the contours of the Maelbeek valley, the quartier Nord-Est was designed in the shape of a truncated diamond with an east-west orientation. Its ground plan resembled an irregular starburst centered on the square Ambiorix—the middle in a string of three squares dissecting the quartier Nord-Est into two equal halves.

The dominant building type was the three-story neoclassical or eclectic rowhouse with rear garden situated on a narrow lot. The convergence of gardens in the interior of street blocks made for a substantial amount of private green space with positive impact on aesthetics and the environment. More substantial residences, some of them mansions, lined the rues de la Loi and Belliard and a few blocks on the western side of the quartier Léopold. Large space users included the Convent of the Dames de Berlaymont (demolished—current location of the EU Berlaymont headquarters) and the Abbey van Maerlant (currently abandoned but integrated into future EU plans for expansion). A small number of mid-rise modern buildings had replaced some of the large mansions in the western part of the quartier Léopold bordering the avenue des Arts (fig. 50).

From the foundation of the quartiers until 1966, land use was predominantly residential. A 1910 map of industrial activity in Brussels and its environs locates only two considerable commercial establishments in the quartier: the maker of luxury carriages Snutzel Frères on the rue de l'Activité, the current rue Jacques de Lalaing, and the furniture-storage facility of the Office des Propriétaires realty on the rue Wiertz.[20] Economic activity in the quartier principally involved small-scale commerce.[21] Industrial activity was concentrated in the

---

19. Institut Géographique National, air photographs of the quartiers Léopold and Nord-Est (detail), sheet nos. 31/467 and 31/201, 1950 and 1956, respectively.

20. Auguste Verwest, *Nouveau plan de Bruxelles industriel avec ses environs* (Brussels: Plans industriels de Belgique, Khiat & Co., 1910).

21. Chambre de Commerce de Bruxelles, *Annales* (Brussels: CCB, 1955).

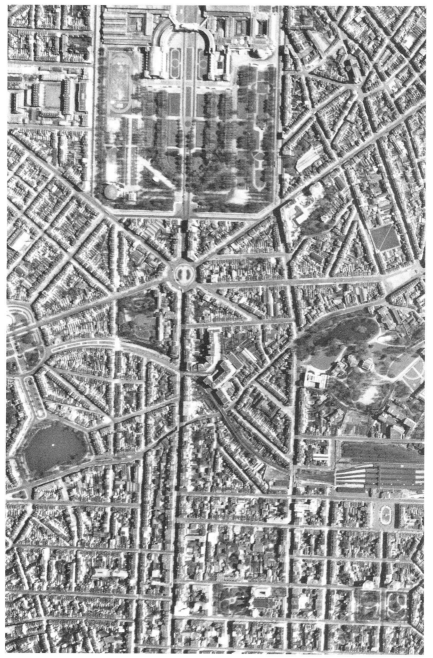

Fig. 48. Air photograph of the quartier, 1950. © National Geographical Institute, Brussels.

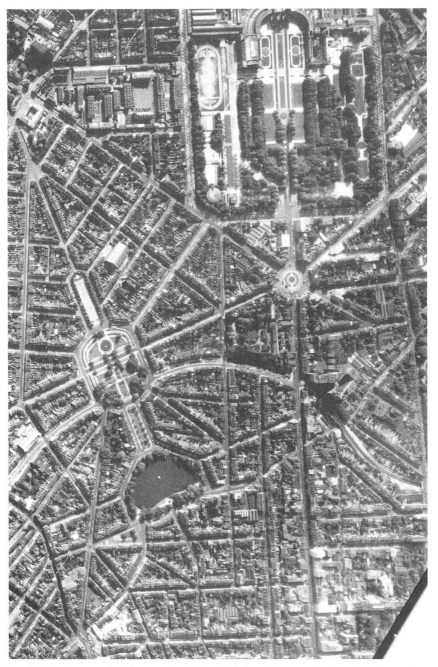

Fig. 49. Air photograph of the quartier, 1956. © National Geographical Institute, Brussels.

Fig. 50. Two generations of office construction. *Top:* Currently vacant office buildings from the 1960s on the rue de la Loi. *Bottom:* Newly constructed mid-rise on the rue Joseph II.

west and northwest parts of the city, specifically along the Senne river and the canal of Willebroek.[22]

The catalyst for change was the decision to place the European Communities institutions in Brussels. Political and symbolic considerations notwithstanding, Brussels and the quartier in question compared extremely well with other sites in terms of centrality, cost of establishment, and management and availability of facilities and services. In a 1967 study published in the *Harvard Business Review*, Brussels scored higher than Zurich, London, Geneva, and Paris as an appropriate site for international business and administrative operations.[23] By 1966, the two main EU headquarters buildings had been constructed and the process of transforming the quartier into an international administrative/business park had been launched.

The Consolidation of the Office Function (1967–85)

In an air photograph of the quartier dating from 1969, it is clear that these changes are already beginning to have a landscape impact on the quartier (fig. 51):[24] The nineteenth-century ground plan remains largely intact, but the first evidence of road modernization has appeared: tunnels have been added to the avenue des Arts, in the perimeter of the Pentagon, and the rue de la Loi as part of a grand infrastructure plan to turn Brussels into an automobile-friendly city: a fitting objective for a city flirting with American urban planning.

At the level of the street block the most patent morphological transformation involves the consolidation of narrow lots into large-size lots capable of accommodating multistory buildings. The Triangle Building on the Schuman roundabout housing the EU translation service was 80 percent completed by this time. Rowhouses still marked the corner of Cortenberg-Joyeuse Entrée. The Triangle made the corner of Loi-Cortenberg the fifth of a total of six intersections facing the Schuman roundabout that have been cleared or constructed upon since 1956 (figs. 52–53).

Further west on the rue de la Loi, rowhouses and mansions were being cleared to make place for mid-rise office buildings. The area

22. *Bruxelles: Un canal, des usines et des hommes* (Brussels: La Fonderie, a.s.b.l., Les cahiers de La Fonderie 1, 1986), various entries.

23. Charles R. Williams, "Regional Management Overseas," *Harvard Business Review* (January-February 1967): 87–91.

24. Institut Géographique National, air photograph of the quartiers Léopold and Nord-Est (detail), sheet no. 31/1313 , 1969).

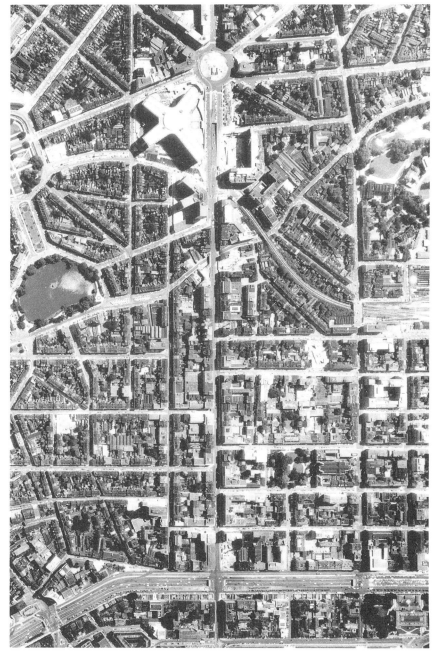

Fig. 51. Air photograph of the quartier, 1969. © National Geographical Institute, Brussels.

Fig. 52. Building activity, 1956–69

rue Charles Martel

rue Franklin

rue Stévin

boulevard Charlemagne

rue Archimède

avenue de Cortenberg

rue du Taciturne

avenue de la Joyeuse Entrée

Rond-Point
Schuman

avenue d'Auderghem

rue Froissart

rue Breidel

Pascale

rue van Maerlant

Park
Léopold

Borchette

Lot under transformation:
Demolition/cleared lot/
building under construction

Data drawn from comparisons of
air photographs of the quartier
(1950, 1956, and 1969).

Source:
Institut Géographique National

Fig. 53. Building activity, 1969–78

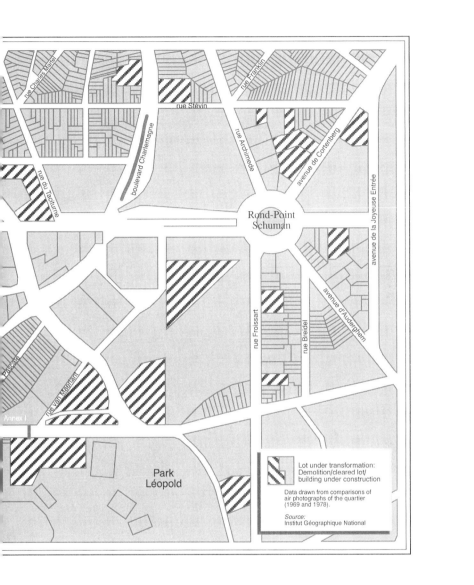

rue Charles Merten

rue Franklin

rue Stévin

boulevard Charlemagne

rue Archimède

avenue de Cortenberg

rue du Taciturne

avenue de la Joyeuse Entrée

Rond-Point
Schuman

rue Froissart

rue Breidel

avenue d'Auderghem

Pascale

rue van Maerlant

Annex I

Park
Léopold

Lot under transformation:
Demolition/cleared lot/
building under construction

Data drawn from comparisons of
air photographs of the quartier
(1969 and 1978).

Source:
Institut Géographique National

between the north side of the rue de la Loi and the rue de Luxembourg was being changed the most. The line of houses separating the art deco Residence Palace from rue de la Loi had been removed. Every block in this area appears affected, with the exception of an irregular rectangular area delimited by the rue de Trèves to the west, the rue Jacques de Lalaing and the massif of the Residence Palace to the north, the parc du Cinquantenaire to the east, and the rue Belliard to the south. The morphology in this area appears essentially unaffected. Modern mid-rise buildings were exceptional here.

The unsystematic manner in which the quartier has been modernized is also evident at the microscale level of the street block. In a 1978 air photograph of the quartier, no blocks appear to have been cleared in a wholesale manner.[25] Targeted street blocks are transformed very slowly: in the beginning, way is usually made for one or two office buildings. A cluster of old housing is demolished in the affected street block and rebuilt as office space, leaving the remaining old buildings bookended by modern mid-rise or high-rise office buildings, and the architectural character of the block and the street diluted. Later, the remaining old buildings in the street block may be removed more readily, since their contribution to forming a substantial aesthetic amenity is deemed severely diminished.[26] This process appears to be important. The modernization is gradual, lacks planning coherence, and is undertaken by a variety of private-sector initiators (fig. 54).

To the north of the Schuman roundabout changes are sparse. High-rise apartment buildings occupying a portion of the built line around the squares Marguerite, Ambiorix, and Marie-Louise are replacing high-quality rowhouses and mansions in the eclectic style of the turn of the century (fig. 55). Otherwise, the dominant building type remains the rowhouse. The condition of the housing stock is quite varied. With its pre–First World War heyday long past, the quartier is generally in a state of benign neglect. Some of the rowhouses are vacant and in great disrepair.

---

25. Exceptions are the blocks on which now stand the Charlemagne and the Berlaymont buildings.

26. In 1990, the few remaining *maisons bourgeoises* from the mid-nineteenth century on the rue Guimard were threatened with demolition or the transformation of their façades by their corporate owners. The argument of diminished aesthetic role in the street was used to secure the permit for their transformation. Finally, a compromise was reached, by which the façades were preserved, and the extension in the form of additional floors desired by the owners was recessed and rendered less visible from street level.

Fig. 54. Rowhouses on rue de Trèves. The rowhouse on the left serves as residential and small commercial space. The rowhouse on the right is vacant and in disrepair.

Land use has broken away from past patterns. The *carroseries de luxe* Snutzel Frères is long gone and the residential function is in full retreat throughout the quartier Léopold. The mid-rise and high-rise office buildings replacing rowhouses and mansions are in almost every case introducing office functions to the area. Certain events shape the course of the quartier's market and the consolidation of the office function. By 1970, a number of large British urban development firms had become active in the real-estate market of the quartier Léopold. This coincides with a period when the city government granted building permits rather easily. Subsequently, during the 1973 oil shortages, the Brussels real-estate market experienced a protracted crisis which brought land prices down significantly. By 1979, the full impacts of the crisis and overbuilding become clear: the quartier has a 20 percent vacancy rate. Following the worldwide recession of the early 1980s, however, investors showed new interest in the quartier (1985 to date).[27]

During this time, EC membership grew from six to nine, ten, and finally to twelve with the accession of Spain and Portugal in 1986. The need for administrative space had been increasing and the EC was then consolidating its presence around the Schuman roundabout, on the rue Archimède and various locations in the old quartier Léopold, such as rue Arlon, the rue de la Loi and the rue de Trèves.[28] Yet, there was little excitement about European unity. The process appeared to have stalled, and the quartier had yet to become the hotbed of European lobbying and business activity it would become after the launching of the "1992" campaign.

The Relaunching of the Community (1986–93)

By 1986 two major events provided the impetus for increased investment in the quartier: the effects of the worldwide recession of 1982–83 had been overcome, and the European Community—by then twelve-members strong—began to consider the Single European Act as the political foundation for the creation of a single unified European market of 345 million consumers. The architect was Jacques Delors, once member of the government of François Mitterand, and the

---

27. Address by the president of the Confédération des Immobilières de Belgique, during the opening ceremony of its national congress in Leuven, 13 September 1991.

28. "Répertoire Télécom," internal document of the Commission des C.E., Brussels, February 1985.

Fig. 55. Intensification of residential land use on the square Ambiorix

newly elected president of the European Commission. With the signing of the Single European Act in 1985, the "1992" era of the quartier dawned.

With land prices skyrocketing in New York, London, and Paris, institutional investors and independents turned an appreciative eye on the markedly undervalued Brussels market. Starting in 1987, Swedish, Norwegian, and Finnish investors flocked to the Brussels land and securities markets along with buying a number of well-situated foreign firms throughout Belgium.[29] Explaining their investment strategy, they evoked the significance of the unified European market and the increasingly international and liberal character of their countries' economies.[30] Most important, however, was the explosion between 1987 and 1989 of real-estate prices in Sweden and Finland, brought about by the liberalization of the real-estate-generated returns to capital of pension funds and insurance companies. The speculative doubling and tripling of land prices and the eventual suppression of returns to capital in Sweden and Finland caused many institutional investors to look outside Sweden and Finland to place their capital. Brussels was one of the top choices. By 1989, Swedish real-estate investment in Brussels alone rose to 30 billion Belgian francs (approximately U.S.$800 million). The Swedish Group Aranäs controls a portfolio worth 15 billion Belgian francs, representing some 150,000 m^2 of office and residential space. Newcomers like Aranäs are delegating the running of their portfolios to long-established urban development and management firms like Jones Lang Wootton and Richard Ellis.[31]

While the Scandinavians have not been unique in their discovery of urban bargains in Brussels, they exemplify the new type of foreign investor. Scandinavian investments in the real-estate sector have been of a character different from that of the British urban development firms of the 1970s: they have been less speculative and have had a longer-term profit horizon. The alliance between Scandinavian investors and British development and management firms is certain to stall any major rethinking of land management or marketing in the quartier Européen-Léopold (figs. 56–57).

The urban morphology of the quartier in this third time frame,

---

29. Anne Vincent, "Les investissements nordiques en Belgique," *Courrier Hébdomadaire du CRISP* [Centre de recherche et d'information socio-politique], nos. 1246–47 (1989): 43–48.

30. Ibid, pp. 3–4.

31. Ibid., pp. 48–50.

1985, exhibits continuity with the past in the intensification of automobile access, the growth of the tertiary and quaternary functions, and the absorption and conversion of the nineteenth-century city. However, it has evolved in three very significant ways.

First, while the first and second time frames saw the consolidation of small contiguous lots into parcels of appreciable size able to accommodate multistory buildings within unaltered street blocks, in this third period we observe the consolidation of street blocks into terrains able to accommodate "megaprojects." The first such project is the new headquarters building of the EU Council of Ministers situated on the consolidated street blocks of rues Juste-Lipse and Froissart. The rues Juste-Lipse and Comines have been suppressed to accommodate the massif of the building (fig. 58). The second project is the Centre International du Congrés (currently leased by the European Parliament), which has absorbed the street block once taken by the Brasserie Léopold and portions of the rues Rémorqueur and Wiertz, and which blocks the north access to the rue Vautier and may expand over the rail yard of the gare du Luxembourg (fig. 59).

The second new feature in the quartier's urban morphology is the systematic abandonment and replacement of mid-rise office towers dating from the late 1950s, 1960s, and even the 1970s, which fail to attract the upscale office consumers who increasingly occupy the quartier. Again, this feature underlines the significance of the market in determining the shape of the quartier. Building types and qualities of construction that were deemed quite adequate for the lackluster Brussels market of past times are now considered portfolio liabilities. With the quartier Européen-Léopold commanding the highest prices in the Brussels office market, new gleaming constructions of granite, anodized steel, and reflective glass are making their appearance throughout the quartier and even on the boundaries of yet unexplored contiguous areas (the new building of the Banco di Roma on the rue du Marteau is a case in point).

The third morphological change involves residential space. Although the green municipalities east and southeast of the quartier continue to attract most of the EU executives and affluent foreigners and their families, gentrification is making significant inroads in the remaining residential portion of the quartier, which largely coincides with the nineteenth-century built fabric. While the rowhouses and residential land use appear quite safe there for the moment, the demographic profile of the area is changing. In the way that Western Europeans and young, frequently single Eurocrats took over the squares at the time when the office function in the quartier was being

Fig. 56. Building activity, 1978–89

rue Charles Martel
rue Franklin
rue Stévin
rue Archimède
boulevard Charlemagne
avenue de Cortenberg
rue du Taciturne
avenue de la Joyeuse Entrée
Rond-Point Schuman
rue Froissart
rue Breidel
avenue d'Auderghem
rue Van Maerlant
Park Léopold

Lot under transformation:
Demolition/cleared lot/
building under construction

Data drawn from comparisons of
air photographs of the quartier
(1978, 1988).

Source:
Institut Géographique National

Fig. 57. Building activity, 1989–91

Lot under transformation:
Demolition/cleared lot/
building under construction

Data drawn from comparisons of
air photographs of the quartier
(1989, 1991).

Source:
Institut Géographique National

Fig. 58. The new EU Council of Ministers complex. *Top:* The construction site as seen from the rue Belliard. *Bottom:* The abandoned Abbey van Maerlant will be converted into an EU information center and reception facility.

Fig. 59. Different views of the Centre International du Congrés. *Top:* View from the rue Wiertz. *Bottom:* View from the place Jourdan.

consolidated, they and their families are now expanding northward in ever greater numbers, pushing the Maghrebin and Turkish working-class population farther and farther north. The housing stock on streets leading north of the three art nouveau squares is sold gradually—at times speculatively and in a relatively short period of time—and is renovated. With the exception of a few new constructions on the boulevard Clovis between the rue des Gravelines and the chaussée de Louvain, most urban investment is limited to the modernization of interiors and the cleaning or restoration of rowhouse façades. If this expansion continues, the chaussée de Louvain will become the new boundary between a largely affluent European population living in amenity-rich rowhouses to the south, and a working-class or welfare-dependent extra-European population living in great densities in deteriorating turn-of-the-century housing to the north.[32]

This third time frame brings us to the present and to historic consideration of the Treaty of Maastricht. Brussels has joined the small club of world-class cities in advanced economies which have a leading role in managing the world economy. The quartier Européen-Léopold is Brussels's business cockpit.[33] Still, the urban morphology of the quartier Européen-Léopold continues to be rather pedestrian in spite of the erection of megabuildings for the Council of Ministers and the European Parliament. The quartier has been marked by a lack of grand planning vision and architectural distinction. A solution to this problem may be found across the rapidly fading Franco-Belgian border.

Building for Mammon or for Europe? The Question of Monumentality

Monumentality as an aesthetic consideration with political implications and its application to the quartier Européen-Léopold need to be addressed. There is no better culture to turn to for a model than the French: writing in 1803, Vivant Denon supported throwing caution to the wind and building on a massive scale: "It is desirable that the greatly significant events of our era be commemorated by colossal monuments."[34] While the negotiation of European agricultural subsi-

---

32. This ghettoization of the immigrant population is a growing concern for the city, but will not be dealt with in this study.

33. This general definition of world-class status is based on King's *Global Cities*.

34. "Il est à désirer que les gigantesques circonstances dans lesquelles nous vivons

dies fails to evoke the romanticism and grandeur of the Napoleonic campaigns, European integration is certainly of comparable significance.

The concept of monumentality has had a variety of incarnations since antiquity: private versus public, sacred versus profane, prosaic versus heroic, literal versus allegorical, intended versus evolved. The physical, literary, and narrative iconography of monumentality has likewise been richly varied in the European tradition. This is not to say that there are not physical—as opposed to literary—monuments which perform a narrative function. Roman official art as expressed in the column of Trajan and the great number of triumphal arches had as primary role political propagandism: the pictorial recounting of victorious campaigns was a fundamental feature of this kind of narrative monument. Another kind of physical monument of a narrative character would include those which allow their "reading" as part of a landscape. Pyramids, tholos-tombs, citadels, cathedrals, statues, and even lavish private residences have been and are being erected to preserve the memory of a person, an event, an ideology, or an era.[35]

Discussing the function and aesthetic of the monument in Nis commemorating the massacre of Serbs by the Turks in 1809, Jacques Guillerme says: "[E]very monument is an official relic of the memorable, subject to all kinds of degradation. It is an artifact which, all at the same time, 'proclaims' and grows old; an artifact which age condemns to misunderstanding, oblivion, or in the best case, myth."[36] He traces the appearance of the adjective "monumental" from the simple definition by J. F. Blondel as "tout édifice remarquable" (1771), to the end of the eighteenth century when one can discern in France a way of "naturalizing" the monumental into the aesthetic sensibilities of the culture as part of a two way process: on one hand, a consideration of volume and visual impact; and on the other hand,

---

soient consacrées par des monument colossaux." Vivant Denon, *Discours sur les monuments d'antiquité d'Italie* (Paris: 8 vendémiaire an XII), p. 7.

35. The builder of the Château de Chenonceau, a tax collector for François I, satisfied with the grand character of his new residence, placed an inscription over one of the fireplaces in order to assure himself that he would never be forgotten.

36. "[T]out monument est trace instituée du mémorable, sujette à toute sorte d'érosions. C'est un artifice qui, tout uniment, 'proclame' et vieillit, un artifice que la durée voue à la méconnaissance et à l'oubli; au mieux, au mythe." Avant-propos by Jacques Guillerme, "Monument/Monumentalité: De la plénitude des symboles à la fascination du vide," in Marilu Cantelli, *L'illusion monumentale: Paris, 1872–1936* (Liège: Mardaga, 1991), p. 9.

the imposition of a political character rejecting any sign of the ancien régime. More concretely, he notes that monumentality became an aesthetic preoccupation following the archaeological campaigns of the Institut d'Egypte, and the importation of sketchbooks and accounts of pharaonic art and architecture.[37]

In searching for monumentality in the construction of European space in Brussels, we need first to identify for whom or for what purpose monuments may have been built. Just as the EC administrative park was not founded for the benefit of the urban development sector, neither was the EC founded for the benefit of the French and German energy and industrial sectors. While Altiero Spinelli clearly aspired to a federalist ideal for Europe, Jean Monnet's pragmatism prevailed. The ultimate goal of Monnet's functionalism was not only to make Europeans prosperous, but to give them a stake in peace and cooperation. Wealth and material security were the enticements to cooperation. The creation of the Single European Market of 1 January 1993 has been the pivotal achievement of the Monnet functionalists. The Maastricht Treaty can on the other hand generally be interpreted as Spinelli's and the idealists' day in court.

Is the quartier Européen-Léopold a personal monument to Monnet, Spinelli, and Robert Schuman, the architects of European integration? Is it a monument to European unity and the political ingenuity that helped navigate Western Europe through the uncertain seas of the Cold War, and now perhaps all of Europe? Or is it a monument to the democratic process? Considering *how* the quartier has become Europe's CED, it is certainly not this. Then are there, at least, any urban, architectural, or artistic elements within the quartier that can be considered tributes to the persons, ideals, or events that brought the European community together? With the exception of renaming the Loi roundabout "rond point Schuman," and naming the Brussels-Capital Region's proposed urban plan for the quartier *Espace Bruxelles-Europe*, there is neither toponymic nor specific physical tribute to the European enterprise in Brussels. Perhaps if the treaty base of the EU had made a commitment to Brussels or to any other of the EU centers (Strasbourg or Luxembourg City), a—likely costly—monumental character would have been planned, or provided for, by EU authorities or the host country. The incertitude concerning the permanent siting of the EU institutions is likely partly responsible for the aesthetic poverty of the quartier Européen-Léopold.[38]

---

37. Ibid., p. 10.

38. The provisional status, however, is not an adequate excuse by itself. Although

The office towers occupied by the executive services of the European Union constitute Brussels's only architectural monument to Europe. Unquestionably, the industrial, social, economic, and Regional policies produced in the quartier resonate throughout Europe and shape markets, cities, resource realms, and the natural environment. But should this suffice? Is not the failure to exhibit in architectural form the evolution of the European polity a failure of European architecture and urbanism? Is not the failure to remake the city in the image of the dominant political forces at least a break with Brussels's urban traditions? Or is nearly forty years from the signing of the Treaties of Rome too early to expect urban expressions of these political and economic developments?

It is at least ironic that the mostly extinct nineteenth-century quartier Léopold was designed and launched expeditiously in the 1850s by persons with relatively modest resources who had a then-modern vision for Brussels. Thirty years later Léopold II had already begun planning an ambitious visual closure to that project with the monumental park, arch, and exposition palaces of the parc du Cinquantenaire.[39] These were achievements reflecting the industrial era's urban sensibilities concerning modernization and embellishment and the political symbolism of nationalism, centralism, and empire, even if they were the nearly megalomaniacal vision of a single person.[40]

The arch, park, and exposition palaces of the Cinquantenaire satisfied many functional and aesthetic needs: they provided closure to the rue de la Loi, anticipated its extension as the avenue de Tervueren, and affirmed the symmetry of a series of urban projects conceived in neoclassical aesthetic terms. They provided the context

---

Strasbourg, like Brussels, is a provisional seat of an EU institution, the planning approach has been radically different. The French building program for the European Parliament in Strasbourg has enjoyed significantly better press among architecture critics and planners.

39. The architect Gédéon Bordiau first conceived the various elements of the park in 1872 as extensions of the quartier Nord-Est, which he also designed. Note of Bordiau, dated 5 October 1874 in the Archives Généraux du Royaume, Brussels, Ponts et Chaussées, Batiments, 147.

40. Léopold II, who was clearly preoccupied with posterity, wrote, "Pour être sûr de laisser un nom dans la mémoire des hommes un prince avisé peut choisir: il doit faire la guerre ou élever des monuments." He chose the second venue. His drive to reshape Brussels was so great that when it became clear that the government was not going to appropriate the necessary funds for a triumphal arch of his liking, he bankrolled the project himself through the Congolese Fondation de la Couronne; Liane Ranieri, *Léopold II: urbaniste* (Brussels: Hayez, 1973), pp. 130, 132, 138.

and space for the placement of exposition halls and gardens which, in the words of Leopold II, "contribute to the embellishment of the capital while they become the center for the meeting of people from all social strata."[41] They emulated the monumental Paris of Napoleon III and signified the apotheosis of Leopold II as master builder of Brussels. They provided this nascent national capital with additional ergonomic and aesthetic landscape elements of a West European capital character.

The arch as genre architectural element is significant. The building of arches following the second century CE can be interpreted as a gesture marking a new beginning or territory. Hadrian's Athenian arch bore different inscriptions on its two sides: "City of Theseus" referred to the Athens Hadrian inherited as emperor; and "City of Hadrian" referred to the Athens he built. Triumphal arches in the Roman world were erected as honorific monuments, possibly to commemorate campaigns and military victories.[42] Unlike certain later examples of arches, Roman arches were fully integrated into the architectural layout of cities. The relation of the arch to the surrounding buildings appears to have been lost after the Renaissance. Leon Battista Alberti and Alberti Palladio, among others, reserved a prominent role in town planning for the arch and the arcades gallery or façade. Perspectives closed by triumphal arches were considered the epitome of architectural elegance. The façade of Venice approached by water makes use of this device: The visitor identifies the columns of the Torre di Orologio as the entrance to the square of St.

---

41. "C'est notre désir commun que cet édifice s'achève de façon à contribuer à l'embellissement de la capitale, en même temps qu'il deviendrait un centre de réunion pour les classes populaires." Léopold II, "Embellissement de Bruxelles (1): Discours du Roi à l'inauguration de l'Exposition, 7 juin 1888," Brussels, Archives du Palais royal, Cabinet de Léopold II; in Ranieri, p. 124.

42. There has been a debate about the place of the arch in ancient Roman society. One school of thought supports a religious raison d'être for the triumphal arch , which centers around a rite of purification after the shedding of blood in battle (Noak and Löwy). Another school of thought supports an art historical explanation: starting with the column as base for a single statue, the Romans developed arches as elaborate platforms for larger honorific sculptural compositions, commonly but not necessarily related to a specific military campaign or victory (Nilsson). F. Noak, "Triumph und Triumphbogen," *Vorträge der Bibliothek Warburg* 26 (1925): 147; E. Löwy, "Die Anfäge des Triumphbogens," *Jahrbuch der kunsthistorischen Sammlungen in Wien*, N.F. vol. 2 (1928): 1; Martini P. Nilsson, "Les bases votives à double colonne et l'arc de triomphe," *Bulletin de correspondance hellenique* 49 (1925): 143; Nilsson, "The Origin of the Triumphal Arch," and "The Triumphal Arch and Town Planning," in Martini P. Nilsson, *Opuscula Selecta*, 2 vols. (Lund: Skrifter Utgivna av Svenka Institutet i Athen. CWK Gleerup, 1952), 1: 1003–15, and 2: 1015–1028, respectively.

Mark, and her eye is directed through the piazzetta to the cathedral by the arcades of the doge's palace and the Libreria.

No better modern example of the symbolic and decorative uses of the arch comes to mind than the Arc de Triomphe du Carrousel in the Tuilleries Gardens of the Louvre Palace. A scale replica of the Arch of Septimus Severus at Rome, the Arc du Carrousel was erected by Pierre François Fontaine and Charles Percier to commemorate the victories of Napoleon I in 1805.[43] Baedeker describes it as follows:

> The arch is perforated by three arcades and embellished with corinthian columns with bases and capitals in bronze supporting marble statues representing soldiers of the empire. The marble reliefs on the sides represent: in front, on the right, the Battle of Austerlitz; on the left, the capitulation of Ulm; at the back, on the right, the conclusion of peace at Tilsit; on the left, entry into Munich. On the north end, the entry into Vienna; on the south end, conclusion of peace at Pressburg. The arch was originally crowned with the celebrated ancient *quadriga* from the portal of St. Mark's in Venice, but this was replaced in 1815 by a quadriga designed by Bosio: Triumph of the Restoration.[44]

A similar symbolic function was served by the more massive Arc de triomphe de l'Etoile (1805–36) erected on the same line of sight as the Arc du Carrousel at the end of the avenue des Champs-Elysées. Borrowing liberally from Roman official art, the French Imperium commemorated its grandeur through narrative monuments even in defeat.[45]

The Arc de triomphe du Carrousel, the arc de l'Etoile, and the

---

43. Karl Baedeker, *Paris and Environs with Routes from London to Paris: Handbook for Travellers* (Leipzig: Karl Baedeker, Publisher, 1907), p. 70.

44. Ibid.

45. All of these monuments can be criticized for being derivative or out of scale with their surroundings. David Pinkney calls the Haussmannian planning of the Etoile mechanical and out of all human proportion. He writes: "From ground level the observer has no impression of architectural unity, partly because the curving surface obstructs the view across the vast place and also because the place is far too big to be encompassed in a single view. It is impressive by its size, but aesthetically it is much less satisfying than the more subtly designed Place Vendôme or the Place de la Concorde." David H. Pinkney, *Napoleon III and the Rebuilding of Paris* (Princeton, NJ: Princeton University Press, 1972), p. 217. The same can be directed as a criticism of its modern cousin, the Grande Arche de la Défense. Still, in their failure to please *all* experts, they are still admirable elements of grands ensembles created by a culture enamored with the monumental.

Marble Arch in London are freestanding and isolated from their surroundings. Nineteenth-century arches are derivative of their Roman
counterparts. They provide an architectural historical linkage with
the Roman past, support claims of cultural continuity and superiority,
and affirm the importance of military conquest.

A similar formal language is put to use in Brussels. Léopold II aspired to do for Brussels what the Empire did for Paris, and erected
the Palais de Bruxelles, the arch of Cinquantenaire, the exposition
halls, the Museum of Central Africa, arranged the opening of the
boulevard du Souverain and the avenue de Tervueren, and undertook many other significant projects which he started and never saw
completed, and which were to eradicate what he probably perceived
as the provincialism of Brussels.

This discussion does not intend to romanticize Léopold II's urban
planning activities. While the Léopold II aesthetic was heavy-handed
and perhaps inappropriate—if not outright destructive—for the
Flemish Brussels landscape, it at least embodied a global vision with
both a functional and an aesthetic coherence. Architecture and construction occupied the same side of the battlefield, and the enemy
was medievalism and provincialism. This kind of coherent vision
and systematic application has been missing in Brussels ever since.

This is not to suggest that the aesthetic impoverishment of the
quartier Européen-Léopold can be addressed by the erection of one
or more triumphal arches, although the aesthetic importance of the
nearby Léopoldian arc du Cinquantenaire is clearly undersold by the
authorities and the private developers in the shaping of the quartier.
It is less the triumphal arch than the spirits of excellence, creativity,
and grandeur that are missing from the quartier, and, as Léopold II
proved, the impediment is rarely solely scarcity of capital.

A building project of any proportion brings together a variety of
resources and agents in a regime of cooperation shaped by markets,
capabilities, and tastes. The creative resources of the architect are the
contour of the land, water, building materials, design, the existing
built context, the functional fit, the cultural/historical palette, and the
choice of rejecting this palette. The resources of the contractor are the
architect, the available budget, the initiator, the nature of demand,
and the ability to coordinate resources. The resources of the initiator
are the architect, the contractor, and the idea for the project. The idea
contains the time horizon of the project, the probability of its completion, its function and profitability (monetary or other), its role in a
strategy or grander scheme, and notions relating to its design and
aesthetic.

Some of these elements are intrinsic to the planning and completion of any project. Others may not be appropriate in the consideration of all types of building projects. These elements of the original idea are given different weights by different initiators of projects: the arc du Cinquantenaire was unimportant in itself but crucial to the completion of the quartier Nord-Est and the opening of the avenue de Tervueren. Its utility was hence in part symbolic. The time horizon was the seventy-fifth anniversary of Belgian independence, but it was meant to stand forever. It was never a commodified project. It was a white elephant proposition which burdened the purse of the state and the resources of the monarch, and its profitability was not even a factor in its planning.

Certainly the founders of the modernist school sought to reorder the list of top priorities, regarding function, social responsibility, mass production, and revolutionary design. The modernist movement never made the kinds of inroads in Belgium that it made in Germany or France. Its impact in the 1920s and 1930s remained limited to the rethinking of the cité-jardins and private dwellings.[46] Until the intended completion of the quartier Nord, the Cité Administrative inside the Pentagon remains the largest modernist project in Brussels.[47] It has generally been regarded as a planning failure.[48] Modernism did not produce a Belgian avant-garde tradition that found application in the quartier Européen-Léopold. Nor did its more revolutionary ideas expressed in Le Corbusier's "La Ville contemporaine," and "La Ville radieuse," have any impact in the ordering of building density in the quartier.[49] The development of the

46. Francis Stauven, "L'ideologie du modernisme belge après l'Art Nouveau," *L'architectura in Belgio: 1920–1940*, 32, no. 2, (1979): 6.

47. As Lichtenberger correctly points out, in large European cities there is a reversal of building heights from the center to the periphery out of deference to the preservation of the ancent core of the city. Interestingly, Brussels's Cité Administrative violates this generally correct tenet of European urban planning, by incorporating a skyscraper within the area of the medieval city. More orthodox and consistent to the height reversal tenent, although not unproblematic, was the planning of the quartier Nord immediately to the north of the medieval city. Elisabeth Lichtenberger, "The Changing Nature of European Urbanization," in Brian B. Berry, ed., *Urbanization and Counter-Urbanization* (Urban Affairs Annual Reviews vol. 11) (Beverly Hills: Sage, 1976), p. 95.

48. The vast plazas of the Cité Administrative were supposed to be used by lunching and promenading office workers on their breaks. They are empty at virtually all times of the day and every day of the week. Their various passageways are littered and grafittied, and the greenery sparse and poorly tended.

49. Robert Fishman, *Urban Utopias in the Twentieth Century: Ebenezer Howard, Frank Lloyd Wright, Le Corbusier* (Cambridge, MA: MIT Press, 1977), p. 190. Le Corbusier was, of course, designing the ideal city and not a commissioned quartier. Still, the two basic

quartier has been left to the initiator, and the result has been piece-meal planning, as has been emphasized throughout the present study. While it is difficult to identify a single "idea" held by all initiators of projects in the quartier, it is probable that function and profitability are the determining factors. Aesthetics, global coherence, and an appropriately lengthy time horizon appear to be of secondary importance.

Yet neither the critics not the apologists of the quartier Européen-Léopold can hide behind free market behavior to explain the unsatisfactory condition of the quartier. In no way can one assume that the free market and its agents are wholly responsible for the lack of monumentality of the quartier. In fact, were it not for the positive reappraisal of the quartier by the international business market and the glitter of the single free European market, the second- and third-rate office towers of the sixties and seventies would have remained the building standard. The land market provides needed capital to the quartier and has become an important source of income for the economy of the city. The issue is not that the market should be free and competitive—which is, at least rhetorically in the case of Brussels, a given—but rather that the immense resources it commands should be used to serve the goals of the private sector, the needs of the city, and the material and political—qua symbolic—needs of the European Community in a comprehensive and systematic manner.

Despite their own illustrious traditions as builders, the Belgian government, the quartier's place entrepreneurs, and the popular opposition to the expansion of the office function in the quartier have failed to rise to the occasion offered by the European Communities and build grand-style facilities befitting the importance of European integration. The growth coalition appears invested in the idea that monumentality is contradictory to economic and political expediency. It has consequently settled for a parochial planning and architectural vision for the quartier Européen-Léopold, founded on short-term profit making. The groups opposed to the Brussels growth coalition, the ARAU and IEB, have also failed to grasp the scale and structure of the European operation in Brussels. Their proposed project for the development of the gare Josaphat as an alternative EU "cité" corrects

---

premises of the *villes idéales* he designed—one, "to order is to classify [functions]," and two, the paradoxical "decongest the city centers by increasing their density"—which widely influenced modern city planning, are completely miscast in the quartier Léopold.

some of the externalities of the unbridled expansion of the office sector in the quartier, but fails to allow for the liberal and imaginative expansion and expression of the European enterprise. Moreover, it fails to recognize the quartier as an important CED and attempts to dictate its structure by suggesting building types and land uses which appear unjustifiable given the prior demands for space of executive land uses (for example, a European Hall of Music).

Perhaps the examination of competing aesthetic paradigms, such as La Défense business district in Paris, can serve in the exploration of contemporary urban monumentality.[50] It is important to discern whether monumentality is or is not an obsolete dimension of planning; whether planning at the monumental scale requires an uncommon consideration of costs and benefits; whether *unique* land markets are best served by exceptional architectural and planning vision (what is the likelihood that another capital of the EU will ever again emerge?); whether spatial linkage of the project to its surroundings can benefit rather than hinder the aesthetic and functional calibration of the project; and finally, whether political consensus and strong leadership are fundamental to the successful conclusion of a monumental urban project.

It is equally important to consider whether monumentality is an appropriate design element in highly politicized urban projects such as the EU administrative park. As Peter Davey, editor of the *Architectural Review* notes, "In a plural democracy, there is a commitment to change, to the development of the individual and to variety. Yet we need ways of making buildings in which democratic affairs can be conducted, and they must be reasonably permanent, partly for economy, but more importantly perhaps to give a sense of continuity."[51] He goes on to make the important point that new ways of building for the executive branch need to be devised. These new built forms should make the necessary bureaucracy efficient and accessible to constituents. In other words, transparency in the policy realm should be reflected in how accessible the seat of policy making

---

50. The consideration of the Tête Défense project in Paris is an appropriate monumental counterpoint to that offered by the megaprojects of the quartier Européen-Léopold. The two will be contrasted briefly in the concluding chapter, but the full elaboration of the merits of monumentality in the design of la Tête Défense requires detailed treatment in a different forum. Admittedly, not all see Mitterand's neoimperialist aesthetic as an improvement on monumental Paris. The Great Arch was described by the *Architectural Review* as "a skyscraper in a hoop."

51. Peter Davey, "Building Democracy: Can We Produce Architecture That Is Emblematic of Common Purpose?" *Architectural Review* 193, no. 1153 (March 1993): 17.

is. The Dutch appear to have succeeded in building the newly constructed Chamber of the Netherlands Parliament *(de Tweede Kamer)* just that way, grafting it into the historic city in a way that allows contact between citizens and politicians. Peter Blundell Jones gives further substance to the argument in his article on Bonn's federal administrative park. The Bonn park is distinguished by its restrained modernism expressed in low-rise glass pavilions in the Miesian tradition, and its deference to river views and the older built context.[52] Transparent buildings for a transparent government, perhaps? It is in many ways the antithesis of the Brussels EU park and the EU bureaucracy.

In previous chapters we discussed the importance of agency in the guise of key political figures and business persons in shaping economic choice and political decisions concerning the quartier. We also discussed the importance of structure in the guise of planning regulations, shifting global market forces, the agenda of the European bureaucracy, and treaty environment in bringing into existence a specialized central business district. We have attempted to combine the two in a working framework which recommends that regimes of cooperation are the key device by which individuals manipulate structures for their own benefit or to some specific purpose. The resulting process (in this case, the process which gave rise to a CED) appears economic and mechanistic. Yet the object of investigation— the quartier—has grown like a live coral reef on a sunken nineteenth-century cutter. Its raison d'être—at least the intended one— was not economistic.

While the role of this study is not prescriptive, the discussion of the processes that shape the quartier Européen-Léopold presents an irresistible opportunity to challenge the limited vision of its shapers. The hope is that such criticism will start a debate about the appropriateness of a more original planning and architectural character for this quartier of a capital importance.

---

52. Peter Blundell Jones, "Bonn, the Provisional Capital," *Architectural Review* 193, no. 1153 (March 1993): 18–33.

9

Conclusions

WOLFGANG BRAUNFELS arrived at the conclusion that cities and city plans that have proved most enduring were not built for strangers but for and by people who lived in them and were being shaped by them. The cities did not anticipate history but rather became the theater for it, and they shone especially brightly when urban development went beyond the mundane fulfillment of functional needs.[1] This pronouncement bears a special significance and warning for Brussels: the quartier Européen-Léopold is essentially being built for strangers, by people who do not live within it and have not had the opportunity, or sobering necessity, to be influenced by local conditions. European history being written within its limits, by contrast, communicates a raison d'être of monumental significance. Indeed, the quartier has unspoken pretensions of becoming a superspecialized central business district—the first true central executive district—at the service of the European integration enterprise. The new morphological character of the quartier—essentially a new quartier created from parts of other quartiers—is quite recent and is based on such pretensions. The introduction of the office function in the eastern fringe of the quartier Léopold dates from the late 1950s, and the first European Community building was constructed only in 1965. In typical Brussels fashion, however, the urban impact of this process of internationalization has been rapid and overwhelming.

The Findings

This study has examined the planning, economic, political, and aesthetic processes and agents transforming Brussels into a world-class

---

1. Wolfgang Braunfels, *Urban Design in Western Europe: Regime and Architecture, 900–1900* (Chicago: University of Chicago Press, 1988), pp. 368–71.

administrative and business capital. The studied processes and agents contribute to the understanding of how the nineteenth-century quiescent residential quartiers Léopold and Nord-Est became the home of a special-purpose CBD—a central executive district—describe the nature of this new urban morphological type, and anticipate that CEDs may become more frequent among world-class cities, as economic but also political functions become increasingly global in character.

The area of study is labeled, for the purposes of this study, the "quartier Européen-Léopold." The selection of this name is of some significance to the object of the study. The quartier is "Européen" for being, on the one hand, the site of the installations of the executive of the European Communities, and on the other hand, the site of a superspecialized executive area—a CED—with distinct referents to the process of European integration.

The name of the quartier also needs to evoke the nineteenth-century past—hence the name "Léopold" after King Léopold I, both because of its urban planning pedigree and because this past still constitutes an active built matrix which influences market choices, molds political realignments among government and business elites, as well as constituencies of resistance, and shapes aesthetic choices and architectural sensibilities. Lastly, the use of French for the quartier's name is appropriate because of what has been described in this study as an overwhelming influence by nineteenth-century French urban planning sensibilities in Brussels during the era of nationalism and the Industrial Revolution.

The balance between the two components in the quartier's name is in flux, as is the quartier's urban morphology. The Léopold component is clearly waning in material significance, as the process of urban transformation has—until recently—been relentlessly consuming the relic built environment and producing permutations of modernist low-rise and mid-rise office towers. The relatively recent rallying of civic resources and organizations against the unchecked demolition or transformation of nineteenth-century buildings has neither produced the necessary political consensus, nor materialized in a timely enough fashion, to make a significant difference in preservation. It is proper to speculate that in a relatively short time in the future, the fitting name for this particular quartier of Brussels will be "quartier Européen."

The method of investigation relied on four interdependent processes behind succession in land use, building types, management control, and change in the functional character of the quartier.

These are (1) a planning tradition that sanctions the idiosyncratic development of the quartier, (2) a land market that is equally firmly linked to the international capital market and to local business interests, (3) a political process that gives rise to ephemeral local and international regimes of cooperation between political and economic elites, and (4) an aesthetic vision and vocabulary dominated by market considerations, the commodification of architecture, and the rejection of Western traditions of urban monumentality.

The underlying tension between, on the one hand, structures, such as planning and the world capital market, and, on the other hand, agency, as embodied by individuals and groups operating in the quartier, is of fundamental importance to the method and the explanation. Much of the literature emphasizes one or the other as the basis of urban transformation. The study advances the notion that structures and agency are causal partners in the transforming of the quartier. One of the key tasks of this study was to find a methodological means that would allow the smooth inclusion of both in the explanation. The selection of the above processes was made with the nature of causality in mind. These processes are labeled "interdependent" as a shorthand for the way structures and agents are married in the centerpiece methodological notion of the study, the regimes of cooperation.

Each of these factors—the planning dimension, the land market, the political regime formation, and the aesthetic vision—deserves a brief recapitulation and summation of related findings.

The Planning Dimension

Brussels is an atypical Western European city because of the historical incidents that made it a capital city, its peculiar urban regulatory tradition—or lack thereof—and its recurrent flirtation with American-style urban development practices and planning. The planning dimension cannot be simply subsumed under the rubric of "political activity," although it is the product of political bargaining between political authorities, business, and citizen groups. Planning regulation, at least in the case of Brussels, exhibits great durability, perhaps because its significance lies less in politically negotiated regulations and more in a certain esprit that prescribes how far political forces can go in tampering with the laissez-faire climate of the city.

The atypical character of the city is captured in the role the private sector has played in shaping planning decisions. This role has been basic to Brussels urbanism. Since Brussels became the capital of the Kingdom of Belgium in 1830, finance capital and the private sector, as represented by the landed aristocracy and prosperous bourgeoisie, have been key participants in the modernization, expansion, and embellishment of the city. The historic opportunity presented to the French-culture elites and the new Belgian and foreign financiers to transform the modest Flemish city into a national capital in the French style, made possible the launching of private ambitious urban development projects, such as the opening of the quartier Léopold, and the Champs-Elysées-like avenue Louise.

This prominence of private capital as initiators of large-scale projects, and as implicit subcontractors of the government, remained a prominent feature of urban development throughout the nineteenth century and continues to be perhaps the most dominant feature of urban transformation today.

What fostered this propensity for private-sector involvement in the planning affairs of the city appears to be a laissez-faire, or "laissez-bâtir" climate. The apparent proliferation of paper regulations, such as the *plan de secteur* of 1972, the *plans particuliers d'aménagement* (PPA), and the creation of *commissions de concertation* responsible for reviewing urban development projects, appears to have a modest influence in curbing the activities of the private sector.

Real planning authority rests in the design and approval of PPAs, which permit specific changes usually in a few street blocks of area. PPAs turn Brussels urbanism into a palimpsest where, theoretically, specific needs of the city and market are addressed without disrupting the environs of the affected area. Its critics, of course, claim that it has a domino effect around the city: Concentrations of PPAs in certain areas of the city weaken the existing built fabric. This appears to be true in the case of the quartier Européen-Léopold.

The significance of this peculiar planning tradition that overtly relies on the private sector for urban initiatives and allows an appreciable amount of flexibility of response to market needs (the PPAs) is that it provides a smooth terrain for urban development firm operatives and an attractive business arena for international capital markets watchers who have been traditionally attracted to high-yielding, high-security, modestly regulated real-estate markets.

A review of spatial conceptions of the city in general and the quartier in particular that have been advanced by planning think tanks, government agencies, and private developers indicates a great attraction to American-style urban solutions. Brussels and the quartier within it have been designed to be automobile-friendly. The quartier and its eccentric counterparts at Woluwe-Saint-Lambert and the boulevard du Souverain show a distinct tendency toward mono-functionalism, jettisoning at least the residential function, and in the case of the quartier Européen-Léopold, the commercial function as well.

The planning environment, however, presents somewhat of a paradox. One would expect that the "laissez-batîr" climate would give nearly complete freedom to urban developers in the types of buildings that they construct. By contrast, there is a marked homogeneity in the form and density of the office space produced in the quartier. While explanations can be put forth, they probably do not lie in the planning regulatory environment. The most likely explanation, which is consistent with the notion of private-sector dominance, lies in the nature of the land market, discussed below.

The planning environment provides structural conditions for the flourishing of planning practices compatible with the fast-changing needs of the private sector at both the local and the international levels. It is difficult to discern, however, explicit dimensions in Brussels's planning environment that would have produced a central executive district rather than a garden-variety central business district. Consequently, the planning environment plays a key role in the shaping of the quartier into a CED but is not of critical significance when considered independently from other factors.

The Land Market Dimension

The relationship between the EC and the private sector is evolving continuously in response to the maturation of European Community institutions, power shifts between the main trading blocks, structural changes in the international economy, and most important, the changing role of firms as mediators of international relations. These key changes signify a new type of power-sharing between transnational organizations like the EC and international capital—a regime of cooperation traditionally identified with the sovereign state. The quartier Européen-Léopold flourishes as a result of this new power arrangement, and its land-use character underlines the significance of international political brokerage for international business. The monopoly of European Community executive decision making by Brussels makes the quartier a testing ground for this new regime. [2]

---

2. In terms of clout in the financial markets, Brussels trails New York, London,

This new dynamic and multinodal world economy finds expression in the massive, rapid, and often computer-programmed transfers of capital between markets. The capital markets employ rankings of desirability that distinguish markets as privileged or underprivileged, secure or risky. Financial officers then proceed with investments that match their client's level of desired returns to capital and his or her willingness to assume risk. The securitization of real-estate investment has made the privileged and relatively safe land markets of Paris, London, and Brussels common destinations for capital that may be produced in Scandinavia, the United States, the oil-producing Arab world, or Japan.

The linking of flexible investing to international real-estate marketing in London, Brussels, or any city produces a paradox. The contradiction involves the very concept of flexibility of movement sought by international portfolio managers and the largely permanent storing of capital in buildings by urban developers. There is very little flexibility in real-estate markets—certainly far less than in the securities market. Hence, in terms of economic rationality, there must exist a link or device that allows the process of international investment to work efficiently in local markets. Recent fascination with the international processes of accumulation and the links between the international capital markets and local land markets has overshadowed the critical significance of local entrepreneurship and local elites in *shaping, cushioning,* and *supplanting* international investment activity. It is in this arena that local, national, and international elites and markets meet and interact.

The quartier Européen-Léopold is such an arena. The rapid development of the office function reflects the coupling of international and local capital with elite individuals that act as catalysts for the efficient application of investment—at least efficient in terms of the elites' interests. Boosterism sustains the interest of international capital markets in the city and the quartier, rallies investors and resources of all magnitudes and at all geographical scales, and, most important, nurtures the relationship between the European Communities, the Belgian state, and the city of Brussels. The result is a political economic ecology of local and international processes specific to a special type of urban development—the new, superspecialized "European" CED.

Tokyo, and Paris, all centers of great national power strategically linked to the world economy. Brussels's power, however, does not lie in the sovereign state, but in the challenge of it.

Unlike cities like Chicago, which have a clearly definable central business district and a number of more recent suburban business districts that concentrate the bulk of administrative/office space, Brussels has a number of office districts within the city and lately in its periphery. In simple terms, its CBD is qualitatively made up of a number of fragments, and I would assert that the city now also contains a central executive district that is increasingly divorced from the types of business activity the rest of the city, its CBD, and its eccentric business centers engage in. The CED's linkages are with national and Regional capitals and the highest echelons of their administration and diplomacy; with proxy organizations for these sovereign, or state-like entities; with special-interest high-level bureaucracies, such as the Union of European Employers' Federations (UNICE); and with firms that engage in servicing the transaction needs of these administrative players. Conspicuously absent are commercial activities of any magnitude, a "bright lights" district, and civic amenities such as museums or concert halls—all fixtures in the typical CBD.

The symbiotic relationship between the planning environment and the land market is clear. In purely market terms, we suggest that the emergence of new complexes of administrative activity have created the demand for a new type of "command, control, and communications" center, which markets have rallied to meet. The market has built on the earlier decision of the Belgian government to locate the communities in the residential quartiers Léopold and Nord-Est by gradually transforming them into a central executive district that meets this emerging functional demand. Whereas the explanation is well informed by the investigation of the quartier's land market, there is still the question of how planning and economic choice in the quartier are engineered and set in motion. It has already been suggested that elites play a significant part in this process.

The Political Dimension

The study suggests that the efficient implementation of the urban morphological transformations of the quartier are tied to the ability of political elites of national and Regional caliber to work with business elites of Belgian and French, and to a lesser extent British, origin. Only when such alliances, or "regimes of cooperation," are forged, can this central executive district expand physically and intensify its spatial linkages.

The whole notion of "regimes of cooperation" as the meeting

ground of agents and structures is not new to the social sciences, although it is new to geography.[3] This notion emerged as a school of thought in international relations, although here, the agents are urban political and business elites.

The device of regimes allows the consideration of "ephemeral," "special-purpose," and virtually always "informal" agreements between agents, without excluding the influence of structures, such as complex market forces, and especially the internationalized securities market as it relates to real-estate development. In effect, the political and business elites that operate at the quartier Européen-Léopold come together to accommodate the space needs of the European Communities, which should be viewed as both client and invisible participant in the decision making concerning the material transformation of the quartier.

The European Community is not allowed by treaty to own or manage its physical facilities until the time its leadership—or in effect the national governments which make up its leadership in conjunction with the European Commission—decide on a permanent site for the institutions. Brussels is merely the provisional site of the EC executive; hence, the apprehension among national and local political and business elites that Brussels may lose the Communities, should the Communities find a more enticing urban and working environment somewhere else. The quartier Européen-Léopold regimes exist in part to avert such an event of monumental significance for the city of Brussels and Belgium, and potentially the nemesis of many illustrious Belgian political ambitions, and in part to serve short-term political and financial interests of the regime participants.

The construction of the Centre International du Congrés is a clear example of how the private sector, Belgian political elites, and the EC conclave united in informal, special-function, ephemeral regimes of cooperation to achieve urban goals of interest to the regime players. The nature of international financing, and the inclusion of a parliament-like amphitheater with booths for simultaneous interpretation, constitute evidence. It would have been a folly for private-sector interests to build such a massive complex in the hope of attracting the

---

3. In fact, regimes theory, as has been elaborated by international relations theorists, is markedly different in certain key respects from the notion of urban regimes of cooperation represented here, most notably because it does not address the issue of tension between structure and agency. The primary reason for this has been the nature of the regimes that international relations theorists have focused on: mostly large-scale trading regimes, such as the GATT. At that scale of interaction, "agency" is subsumed under the persona of the state.

European Parliament committee and eventually the plenary sessions. It would have been illegal for the EC to give any guarantees to the private sector that this project could be leased by the EC. Finally, it would have been politically risky for Belgian politicians to extend public land and building permits to a project that diverts investment from other areas in the city and especially from housing, which is in such short supply in Brussels, to a "prestige project." However, it makes perfectly good sense for all concerned, involving a relatively modest amount of risk, if a cooperation game or regime was at work here. European parliamentary sources indicate the existence of such a regime, as do the building permits and the corporate composition of the consortium that is building the CIC. The CIC has been, of course, leased by the European Parliament, as feared by its critics and hoped for by developers and certain political elites in the city.

Returning to the fundamental question about structure and agency in the transformation of the quartier, we conclude that structures (such as firms and political institutions) serve as indispensable frames of activity in the design and implementation of Brussels's urban policy, in spite of the fact that individual agency (as exemplified by key persons such as powerful, long-established national and Regional politicians), is essential to the running of the structures. Structures and persons carry out a policy agenda by building regimes of cooperation, which last as long as common interest exists.

We also conclude that these regimes perform spatial tasks by investing the quartier, collections of street blocks, or even single lots with locational significance, which in turn influence future economic and aesthetic choices in their immediate vicinity. Moreover, we have observed that they provide a negotiated code of behavior for regime players that has impacts on the function and intensity of land use. Regimes also appear to broker the supply and demand of both political and economic capital at the local, Regional, national, or transnational levels that again have urban impacts: they contribute capital, add to or alter the functional character of the quartier, and bring sometimes alien planning, technological, architectural, or aesthetic sensibilities to the existing growth culture of the quartier.

The political dimension almost completes the explanation for the emergence of a central executive district at the quartier. Not only the "shape" but also the "shaping" of the CED serves the political and economic interests of the agents. We discern that regimes constitute ephemeral forums for self-interested agents, in which land-use development and management are mediated through market brokerage and political patronage.

The Aesthetic Dimension

The aesthetic dimension, broadly defined, addresses two issues: urban morphological evolution, and monumentality. Explicit to this part of the explanation is the notion that the built environment does not act as passive matrix or canvas for urban transformations initiated by political and economic forces, but itself shapes political, economic, and aesthetic choices. Hence, the built environment and its aesthetic sensibility constitute an active structure.

This part of the explanation distinguishes different time frames of landscape transformation in the quartier that are arguably tied to the fortunes of the EC: we have labeled these time frames the launching of the Communities (1957–66); the consolidation of the office function (1967–85); and the relaunching of Europe (1986 to date).

The mode and scale of transformation changed from time frame to time frame. In the period of the launching of the Communities, when European integration did not captivate the imagination of politicians, electorates, and markets as it does today, investment and change appeared haphazard and limited to the subblock level. The approach to urban morphological change could be described as the "Trojan horse" approach, as a single low-rise office tower would debut in a street block of turn-of-the-century residential rowhouses, only to start a cascade of demolitions of rowhouses in subsequent years. The resulting pattern of "modernization," "intensification," or "Europeanization" of the quartier by the private sector thus appears unordered, in contrast to the apparent desire of the European Communities to concentrate their facilities around a specific roundabout: the rond-point Schuman.

The consolidation of the office function from 1967 to 1985 reflects the cumulative and substantial impact of the gradual, unorderly private-sector development of low-rise and mid-rise office towers. This expansion of office space has in essence removed the greater part of nineteenth-century and early twentieth-century urban fabric of the old quartier Léopold and is now infringing upon other nineteenth-century extensions of the city to the north and to the south. In this period, the scale of projects does not change, but we encounter, on the one hand, the systematic diffusion of administrative activities away from the EC hub of the rond-point Schuman and throughout the quartier (clearly the result of increased space needs of the EC in part due to the accession of new member states), and on the other hand, the expansion of a small cross-section of typical CBD front-office operations. As noted earlier, curiously absent from this CED were commercial activities of any significance.

This development reflects the commitment of the Brussels business community less to the EC, which had still to earn the confidence of the European private sector, than to the quartier Léopold that was emerging as the major node of the multipolar CBD system of Brussels. In fact, the CBD-type office uses proliferating in the quartier in this period did not have a strong referent to the European integration enterprise.

It is the third period, the so-called "relaunching of Europe" following the signing of the Single European Act in 1986, popularly known as "1992," that propelled the quartier out of its relative obscurity as a business land market and international administrative hub and into the political limelight and the securitized international real-estate marketplace.

The changes have now emphatic referents to the European integration enterprise. The scale of the projects changes conspicuously, not any longer limited in extent to parts of street blocks or to the street block level, but absorbing portions of the old ground plan into megaprojects for the consumption of the EC and a vast new array of lobbyists, embassies, national representations to the EC, Regional representations to the EC, and European-level professional, industrial, and labor societies.

The question of monumentality in the quartier—or the lack thereof—has also been raised by the study. The study suggests that one can take issue with the lack of a grand planning vision and architectural distinction in the quartier; incidentally, both of these are listed as major complaints of real-estate developers doing business in the quartier. In spite of these complaints, however, it is clear where the priorities of the parties active in the quartier lie: in making the urban morphological and aesthetic decisions that meet the sophistication and purse size of the clientele. Moreover, the provisional status of Brussels as seat of the EC executive has had a negative impact on the attitude of the key arbiter of planning, the Belgian and Brussels authorities. Thus, the quartier Européen-Léopold CED is not invested with the grand aesthetic one would associate with the long tradition of monumentality in European capitals. Instead, one has to search diligently among the often drab office towers to find either a restored architectural gem postdating the EC activities, or an aesthetically valuable modern or postmodern structure.

~

The Brussels CED is a new urban type that may anticipate the growth pattern, function, and even look of similar districts in a small set of world-class centers that already are or eventually will be beneficiaries of the growing trend in the globalization of command, control, and communication functions associated with administration and corporate activity. It also signifies a permutation of the traditional CBD, the definition of which has not changed notably since it became a feature of the industrial, capitalist city. The study of the quartier Européen-Léopold may stimulate the discussion about the nature of central business districts and the apparent tendency for superspecialization of their functional character in advanced capitalist societies.

Unresolved Problems

An underlying question of importance that the quartier Européen-Léopold study has left unanswered is the role of the democratic process in the shaping of urban places. While the outrage of Georges Timmerman, Paul Staes, David Harvey, Darrel Crilley, and Paul Knox may be justified at some level, since the urban coalitions that created Canary Wharf in London and the quartier Européen-Léopold have been making a mockery of the democratic process, perhaps we should ask what is so novel or different about this undemocratic process. In what ways has it differed from what we have seen taking place in London and Brussels at least since the Industrial Revolution?[4] Is there an inherent instrumental authoritarianism in "cities that work"; and is a democratic city *à la carte* compatible with the imperatives of metropolitan development in the capitalist system?

A second question important to the understanding of the evolution of Brussels's urban political culture involves the apparent break with the past in the manner in which the ethnolinguistic rivalry between Belgium's two major cultural communities—the Flemish and the Walloons—relates to the processes shaping the quartier. Conventional wisdom would suggest that the ethnolinguistic rivalry should qualify the contentious issues concerning the radical urban transformation of the quartier, pitting one community against another in at

---

4. Georges Timmerman is an independent Brussels journalist who has written on the collusion between political and economic elites in the shaping of the quartier Européen-Léopold. Paul Staes is a member of the European Parliament's Green Party, and a fervent opponent of the expansion of the EC administrative park in the quartier.

least some areas of policy. To the extent that this question was pursued by the study, it appears that the rivalry has not played a significant role in the shaping of the quartier, as opposition and support for the changes have cut across ethnic lines. If this is indeed the case, then we are witnessing a breakthrough in communal relations in Brussels. If it is not the case, then it is important to know the ways in which this all-consuming rivalry may be shaping the future of the quartier Européen-Léopold.

A third question involves the apparent shifts in the demographic and human ecological composition of the quartier. All necessary census data up to and including the results of the 1980 census are currently available. The latest census figures should confirm the pressure the expansion of the EU park is imposing on vulnerable populations to the north and south. Unfortunately, the size of Brussels's census tracts (7–8 blocks) is large and helps hide part of the sociodemographic transition. It is especially regrettable that authorities have not redistricted the city into smaller census tracts, as was hoped after the redistricting of the city into 722 tracts in the 1980s. Since the most dramatic urban morphological transformations in the quartier have taken place since 1985, such detailed census figures would have been of critical importance in determining how the quartier has changed as a human community since the "relaunching of Europe." This story somehow needs to be told.

Questions about the impact of European integration on the rest of the city also warrant consideration. This study focused as narrowly as possible on the quartier itself with limited reference to material changes around the Brussels metropolitan area. Attention should be given to the issue of the quartier Européen-Léopold as "growth pole" for the immediate surroundings and for the city as a whole, and to the relationship between competing nodes in the Brussels area: specifically, the relationship between the rapidly expanding quartier Nord, the suburban office developments on the boulevard du Souverain, and the extraurban corporate developments southeast and east of the city. These additional studies will confirm the degree to which physical proximity to the European institutions and the ability of businesses and lobbyists to foster systematic interpersonal relations with EC elites are the key reasons for the specialized functional character of the quartier Européen-Léopold.

Finally, a critical theory not limited to Marxist discourse needs to be developed to address issues relating to megaprojects and to the emergence of international central business districts and central executive districts in world-class cities. The notion of flexible accumulation does not adequately explain the proliferation of urban forms such as the quartier Européen-Léopold, London's Canary Wharf, and Paris's La Défense. This empirical study has attempted to formulate an analytical framework by drawing on land rent, regimes, and urban morphological theory, which could provide a basis for further theoretical work.

Appendix: Building Permit for the Centre International du Congrés

Ministry of Brussels

Town and Country Planning
Administration

REGISTERED MAIL

1000 Brussels
7–9 Rue Ducale

Ministry of Brussels Region
Attn: The State Secretary
Mr. J.L. Thys
7–9 Rue Ducale
1000   BRUSSELS.

Our reference:
43AB/68.633

BUILDING PERMIT

The State Secretary,

having regard to the application submitted on 27 April 1987 by the Minister of the Brussels Region and received on 27 April 1987, relating to a property in Brussels, 43–47 Rue Wiertz, and proposing the demolition of existing building and the construction of an International Conference Centre,

having regard to the organic Law of 29 March 1962 on town and country planning, and in particular Article 48 thereof, as amended by the Law of 22 December 1970,

having regard to the royal decree defining what legal persons governed by public law shall be issued with permits for building and

263

parcelling out land by the delegated government officer, what form the latter's decisions shall take and how applications for permits shall be examined,

having regard to the ministerial decree of 6 February 1971 delegating ministerial powers with regard to town and country planning and designating the delegated government officer,

having regard to the royal decree of 5 November 1979 laying down for the Brussels region specific publicity measures applicable to certain applications for permits for building and parcelling out land and creating for each of the communes of the Brussels region a coordinating committee to deal with local land planning matters,

having regard to the opinion of 16 July 1987 of the College of the Mayor and Aldermen of Brussels,

having regard to the opinion of the coordinating committee of 9 June 1987 and 30 June 1987,

whereas the area where the property is situated is covered by a regional plan issued by royal decree on 28 November 1979,

having regard to the building regulations of the commune and the Brussels Conurbation,

whereas the application contravenes the building regulation of the Brussels Conurbation and whereas the College of the Mayor and Aldermen of Brussels issued a favorable opinion on the application for exemption, through its decision of 16 July 1987,

having regard to the royal decree of 27 April 1987 partially revising the regional plan,

whereas the project is in the public interest,

having regard to the decision of the Brussels Regional Government of 27 April and 8 July 1987,

DECREES

Article 1 - the permit shall be issued to the Ministry of Brussels Region (plans No. 1–24 and amendments No. 17–20 of 29 July)

which must:

1) abide by the conditions of the Fire Service of Brussels Conurbation, as laid down in its notice, a copy of which should be asked for;

2) abide by the requirements of the royal decree of 9 May 1977 implementing the Law of 17 July 1975 on access for the disabled to public buildings;

3) abide by the agreement of 26 June 1987 between the Brussels region and the group of investors, to ensure that the whole site is parcelled out correctly as laid down in urban planning certificate no. 2, the subject of the present decision;

4) obtain formal guarantees ensuring the continued use of the Leopold Park and improved access for the public;

5) The International Conference Centre must have enough parking spaces from the beginning of its entry into service;

6) in view of the depth of the excavations, ensure that the roots of all the trees near the earthworks are surrounded by protective casing before work starts (in cooperation with the Municipal Parks and Playgrounds Service). The entrepreneur shall not move any areas of shrubs without the consent of the above service. Building site vehicles shall not be permitted to drive around in the park.

7) any alterations to the surroundings shall require a separate application for a permit.

Article 2 - Since the edge of the site is in an area listed by royal decree, the present permit shall not take precedence over the Law of 7 August 1931 on monuments and sites

Article 3 - Responsibility for abiding by the present decrees shall be transferred to the applicant, the College of Brussels Conurbation and the College of the Mayor and Aldermen of Brussels.

Article 4 - The permit holder shall notify the College of the Mayor and Aldermen and the official responsible of the start of work or other authorized activity at least a week before starting this work or other activity.

Brussels, 31 July 1987

Secretary of State

J.L. Thys

Copy for the Council of the Brussels Agglomeration

Secretary of State

[signature]

J.L. Thys

Bibliography

Adshead, S. D. *Town Planning and Town Development.* London: Methuen and Co., 1923.

―――. "The Town Planning Conference of the Royal Institute of British Architects." *Town Planning Review* 1 (1910).

Agnew, John, John Mercer, and David Sopher, eds. *The City in Cultural Context.* Boston: Allen & Unwin, 1984.

Alduy, J.-P. "Quarante ans de planification en région Ile de France." *Cahiers de l'Institut d'Aménagement et d'Urbanisme de la Région Ile de France* 70 (1983): 11–85.

Alonso, William. *Location and Land Use: Towards a Theory of Land Rent.* Cambridge, MA: Harvard University Press, 1964.

―――. "A Theory of the Urban Land Market." *Proceedings of the Regional Science Association* 6 (1960): 149–58.

André, R. "Evolution régionale de la population étrangère de Belgique d'un recensement à l'autre 1947–1981." In *Immigrés: Qui dit non à qui?* Brussels: Editions de l'Université de Bruxelles, 1987.

AN-HYP Banque d'Epargne S.A. *Valeur Immobilière, Avril 1991.* Antwerp: AN-HYP S.A., 1991.

―――. *Valeur Immobilière, Avril 1991; Poit de mire: Les maisons urbaines.* Antwerp: AN-HYP S.A., 1991.

―――. *Valeur Immobilière, Avril 1990.* Antwerp: AN-HYP S.A., 1990.

―――. *Valeur Immobilière, Avril 1989.* Antwerp: AN-HYP S.A., 1989.

―――. *Valeur Immobilière, Avril 1988.* Antwerp: AN-HYP S.A., 1988.

Anthony, H. A. "Le Corbusier: His Ideas for Cities." *Journal of the American Institute of Planners* 32 (1966): 279–88.

ARAU. "Le Carrefour de l'Europe de plus en plus malmené." Press conference, Brussels, 14 April 1992.

―――. "Janus place du Luxembourg: Une nouvelle idée de l'ARAU pour sauver vraiment la gare du quartier Léopold." Press release, Brussels, 13 March 1992.

————. "Berlaymont: L'Executif régional et son Président Charles Picqué vont-ils continuer à décevoir la population par leur immobilisme?" Press conference, Brussels, 14 June 1991.

————. "Terminer la reconstruction du carrefour de l'Europe en contrebas de la cathédrale, de la Gare Centrale et de l'Albertine." Press conference, Brussels, 2 May 1991.

————. "Quand le pourrissement délibéré du patrimoine immobilier va de pair avec l'avilissement de l'environnement?" Press conference, Brussels, 30 July 1990.

————. "Continuera-t-on à laisser le secteur immobilier dégrader sciemment l'image de Bruxelles Capitale de l'Europe?" Press release, Brussels, 6 June 1990.

————. "Aménagement urbain: Le général Comte Belliard cerné par les voitures." Press release, Brussels, 19 October 1989.

————. "Espace Léopold manquent toujours la transparence et la volonté politique de prendre en compte l'environnement urbain." Press conference, Brussels, 3 March 1989.

————. "Espace Léopold: Sous le couvert de l'Europe, la démesure de la promotion." Press conference, Brussels, 21 April 1988.

————. "Unifier les institutions qui gouvernent Bruxelles, pour que la ville ait tout pouvoir sur elle-même." Press conference, Brussels, 26 February 1988.

————. "Construction de bureaux: L'ARAU demande un moratorium de 24 mois." Press conference, Brussels, 15 July 1987.

————. "Extensions des Communautés Européennes: Irrégularités en cascade." Press conference, Brussels, 3 May 1984.

ARAU/Commission Française de la Culture. "Un projet culturel pour l'Europe." *Bruxelles vu par ses habitants.* Brussels: ARAU, 1984.

Arcq, Etienne, and Pierre Blaise. "Les groupes de pression patronaux." *Courrier Hebdomadaire du CRISP* [Centre de recherche et d'information socio-politique], no. 1252 (1989).

Aron, Jacque. *La Cambre et l'architecture: Un regard sur le Bauhaus belge.* Liège: Architecture et Recherches/Pierre Mardaga, 1982.

Aspinal, P. J., and J. W. R.Whitehand. "Building Plans: A Major Source for Urban Studies." *Area* 12 (1980): 199–203.

Axelrod, Robert. "The Emergence of Cooperation among Egoists." *American Political Science Review* 75 (1981): 306–18.

————. "Effective Choice in the Prisoner's Dilemma." *Journal of Conflict Resolution* 24 (1980): 3–25.

Baedeker, Karl. *Paris and Environs with Routes from London to Paris: Handbook for Travellers.* Leipzig: Karl Baedeker, Publisher, 1907.

Bamber, M. J., ed. "Property Report Belgium." *Richard Ellis Research* (September 1992): 2.

Banfield, E. C. *The Unheavenly City: The Nature and Future of Our Urban Crisis.* Boston: Little, Brown 1970.

Banham, R. *Theory and Design in the First Machine Age.* London: Architectural Press, 1960.

Barnes, Trevor, and Eric Sheppard. "The Rational Actor in Space and Place: A Re-Evaluation of the Rational Choice Paradigm." Manuscript, 1993.

Bastié, Jean. *Géographie du Grand Paris.* Paris: Masson, 1984.

———. *La croissance de la banlieu Parisienne.* Paris: Presses Universitaires de France, 1964.

Bauman, J. F. "Housing the Urban Poor." *Journal of Urban History* 6 (1980): 211–20.

Beaujeu-Garnier, Jacqueline. *Géographie urbaine.* Paris: Armand Colin, 1980.

Berman, D. S. *Urban Renewal: Bonanza of the Real Estate Business.* Englewood Cliffs, NJ: Prentice Hall, 1969.

Berry, B. J. L. *Comparative Urbanization: Divergent Paths in the Twentieth Century.* New York: St. Martin's Press, 1981.

———, ed. *Urbanization and Counterurbanization* (Urban Affairs Annual Reviews, vol. 11). Beverly Hills: Sage Publishers, 1976.

———. "Cities as Systems within Systems of Cities." *Papers and Proceedings of the Regional Science Association* 13 (1967): 147–63.

Berry, B. J. L., and John D. Kasarda, *Contemporary Urban Ecology.* New York: Macmillan, 1977.

Black, J. T., L. Howland, and S. L. Rogel. *Downtown Retail Development: Conditions for Success and Project Profiles.* Washington: Urban Land Institute, 1983.

Blaise, Pierre, Evelyn Lentzen, and Xavier Mabille. "L'élection régionale bruxelloise du 18 juin 1989." *Courrier Hebdomadaire du CRISP* [Centre de recherche et d'information socio-politique], no. 1243 (1989).

Blowers, A. *The Limits of Power: The Politics of Local Planning Policy.* Oxford: Pergamon, 1980.

Blundell Jones, Peter. "Bonn, the Provisional Capital." *Architectural Review* 193, no. 1153 (March 1993): 18–33.

Boardman, P. *The Worlds of Patrick Geddes: Biologist, Town Planner, Re-educator, Peace Warrior.* London: Routledge and Kegan Paul, 1978.

Bochart, Eugène. *Bruxelles ancien et nouveau: Dictionnaire historique des rues, places, édifices, promenades, etc. de Bruxelles.* Brussels: 1857.

Bogaert-Damin, A. M., and I. Maréchal. *Bruxelles: Développement de l'ensemble urbain 1846–1961: Analyse historique et statistique des recensements.* Namur: Presses Universitaires de Namur, 1978.

Bolan, R. S. Emerging Views of Planning. *Journal of the American Institute of Planners* 33 (1967): 233–45.

Bonefant, P. "Les premiers remparts de Bruxelles." *Annales de la société royale d'archeologie de Bruxelles* 40 (1936): 7–46.

Booth, C. *Improved Means of Locomotion as a First Step Towards the Cure of the Housing Difficulties of London.* London: Macmillan, 1901.

Bordiau. Note dated 5 October 1874. Brussels, Archives Généraux du Royaume, Ponts et Chaussées. Batiments, 147.

Borremans, Dirk. "La création des sociétés régionales d'investissement." *Courrier Hebdomadaire du CRISP* [Centre de recherche et d'information socio-politique], no. 1237 (1989).

Borsi, Franco, and Paolo Portoghesi. *Victor Horta.* Brussels: Vokaer, 1977.

Bourne, Larry, ed. *Internal Structure of the City: Readings on Urban Form, Growth, and Policy.* New York: Oxford University Press, 1982.

Bradley, Owen. "Etre ou ne pas être à Bruxelles?" *L'évenement immobilièr* 72 (December 1992): 7.

Branford, V., and P. Geddes. *The Coming Polity: A Study of Reconstruction (The Making of the Future).* London: Williams and Norgate, 1917.

Braun, Georg, and Frans Hogenberg. Engraved and illuminated map of Brussels, ca. 1572. Knokke, Mappamundi Collection.

Braunfels, Wolfgang. *Urban Design in Western Europe: Regime and Architecture, 900–1900.* Chicago: University of Chicago Press, 1988.

"Brussels Behemoth." *Architectural Review* 193, no. 1154 (April 1993): 15.

Bruxelles art nouveau. Brussels: Archives d'Architecture Moderne, 1988.

Bruxelles: Un canal, des usines et des hommes. Brussels: La Fonderie, a.s.b.l., Les cahiers de La Fonderie 1, 1986.

Burchell, R. W., and G. Sternlieb, eds. *Planning Theory in the 1980s: A Search for Future Directions.* New Brunswick, NJ: Rutgers University, Center for Urban Policy Research, 1978.

Burgess, M. *Federalism and European Union: Political Ideas, Influences, and Strategies in the European Community, 1972–1987.* London: Routledge, 1989.

———, ed. *Federalism and Federation in Western Europe.* London: Croom Helm, 1986.

Butler, S. M. *Enterprise Zones: Greenlining the Inner Cities.* New York: Universe Books, 1981.

Cabinet du Ministre-Président. *Forum: L'Immobilier à Bruxelles.* Brussels: Executif de la Région Bruxelles-Capitale, 1990.

Cantelli, Marilù. *L'illusion monumentale: Paris, 1872–1936.* Liège: Mardaga, 1991.

Cassier, Myriam, Michel De Beule, Alain Forti, and Jacqueline Miller, eds. *Bruxelles: 150 ans de logements ouvriers et sociaux.* Brussels: Les Dossiers Bruxellois, nos. 7–8, December 1989.

Castells, M. *The City and the Grassroots: A Cross-Cultural Theory of Urban Social Movements.* London: Edward Arnold, 1983.

———. *City, Class, and Power.* London: Macmillan, 1978.

———. *The Urban Question: A Marxist Approach.* Cambridge, MA: MIT Press, 1977.

Centre d'études et de recherches urbaines ERU, a.s.b.l. *Rue aux Laines à Bruxelles.* Brussels: Ministère de la Communauté française, Administration du patrimoine culturel, 1980.

CERAU, *Espace Bruxelles-Europe: Phase F rapport final.* Brussels: CERAU pour le Secretariat d'état à la région Bruxelloise, Administration de l'urbanisme et de l'aménagement du territoire, 1987.

Chambre de Commerce de Bruxelles, *Annales.* Brussels: CCB, 1955.

Chapman, S. D., ed. *The History of Working Class Housing.* Newton Abbot: David and Charles, 1971.

Cheshire, P., and D. Hay. *Urban Problems in Europe.* London: Allen and Unwin, 1987.

Cicin-Sain, B. "The Costs and Benefits of Neighborhood Revitalization." In *Urban Revitalization* (Urban Affairs Annual Reviews, no. 18), edited by D. B. Rosenthal. Beverly Hills: Sage, 1980.

Ciucci, G., F. Dal Co, M. Manieri-Elia, and M. Tafuri. *The American City: From the Civil War to the New Deal.* Cambridge, MA: MIT Press, 1979.

Clark, Gordon L. "A Realist Project: *Urban Fortunes: The Political Economy of Place.*" *Urban Geography* 11, no. 2 (1990): 196.

Clarke, Susan E. "'Precious Place: The Local Growth Machine in an Era of Global Restructuring." *Urban Geography* 11, no. 2 (1990): 186–87.

Clavel, P. *The Progressive City: Planning and Participation, 1969–1984.* New Brunswick, NJ: Rutgers University Press, 1986.

Clavel, P., J. Forester, and W. W. Goldsmith, eds. *Urban and Regional Planning in an Age of Austerity.* New York: Pergamon, 1980.

Cluysenaar, J. P. *Plan d'un Palais destiné au roi, à l'industrie et aux arts ainsi que d'un nouveau quartier y faisant suite, situé entre les portes de Louvain et de Namur à Bruxelles.* Brussels, 1842.

Cnudde, Hervé. "Le projet ARAU de construction d'un second pole Européen sur la gare Josaphat pris en compte par les pouvoirs publics." Press release, Brussels, 19 September 1991.

Cohen, S. S., and J. Zusman. *Manufacturing Matters: The Myth of the Post-Industrial Economy.* New York: Basic Books, 1987.

Compagnie Foncière Internationale. Brochure. Brussels: CFI, 1990.

Conzen, M. R. G. "Alnwick, Northumberland: A Study in Town Plan Analysis." *Institute of British Geographers,* publication no. 27 (1960).

Conzen, Michael P. "Town-Plan Analysis in an American Setting: Cadastral Processes in Boston and Omaha, 1630–1930." In *The Built Form of Western Cities: Essays for M. R. G. Conzen on the Occasion of His Eightieth Birthday,* edited by T. R. Slater. Leicester: Leicester University Press, 1990.

Council of Europe. *Historic Town Centers in the Development of Present-Day Towns: Second European Symposium of Historic Towns, Strasbourg, 30 September–2 October, 1976.* Strasbourg: Council of Europe, 1977.

Crédit Communal de Belgique. *Villes en mutation XIXe–XXe siècles: 10e Colloque International, Spa, 2–5 septembre, 1980.* Brussels: Crédit Communal de Belgique, Collection Histoire Pro Civitate, vol. 8, no. 64, 1982.

Crilley, Darrel. "Megastructures and Urban Change: Aesthetics, Ideology, and Design." In *The Restless Urban Landscape,* edited by Paul Knox. Englewood Cliffs, NJ: Prentice Hall, 1993.

Culot, Maurice. *Le siècle de l'éclectisme: Lille 1830–1930.* Brussels: Editions des Archives de l'Architecture Moderne, 1979.

Curl, J. S. *European Cities and Society: A Study of the Influence of Political Change on Town Design.* London: Leonard Hill, 1970.

Danckaert, Lisette. *Bruxelles: Cinq siècles de cartographie.* Tielt: Lannoo, 1989.

Davey, Peter. "Building Democracy: Can We Produce Architecture That Is Emblematic of Common Purpose?" *Architectural Review* 193, no. 1153 (March 1993): 17.

De Bruyker, Philippe. "Bruxelles dans la réforme de l'Etat." *Courrier hébdomadaire du CRISP* [Centre de recherche et d'information socio-politique], 1230–31 (1989).

De Bruyne, F. *Projet d'assainissement de la ville de Bruxelles: Bruxelles à l'abri des inondations, embellissements, création d'un boulevard de ceinture.* Bruxelles, 1865.

de Fer, Nicolas. *Plan du bombardement de Bruxelles par l'armée du Roi le 13, 14, et 15 Aoust 1695.* Brussels: Bibliothèque Royale, Section des cartes et des plans XXXI, 1695).

De Lannoy, Walter, and Christian Kesteloot. "Differenciation résidentielle et processus de ségregation." In *La cité Belge d'aujourd'hui: Quel devenir?* In the 39th annual special issue of *Bulletin trimestriel du Crédit Communal de Belgique,* no. 154 (October 1985).

———. *Sociaal-geografische atlas van Brussel-Hoofdstad* [Social-Geographic Atlas of Brussels-Capital Region]. Antwerp: De Sikkel/De Nederlandsch Boekhandel, 1978.

de Meulenaer, Th. "L'Europe, ça rapporte . . ." *Vlan* 1361 (19 December 1990): 16.

de Ridder, Paul. *Bruxelles: Histoire d'une ville brabançonne.* Trans. Emile Kesteman. Gand: Stichting Mens en Kultuur, 1989.

De Schrevel, E. ed., *Un Statut pour Bruxelles.* Brussels: Centre d'études des institutions politiques, 1988.

De Smedt, P.-A. "Bruxelles entre son avenir international et celui de ses habitants: Exposé du Président de l'Union des Entreprises de Bruxelles." *Forum Immobilier* (15 and 16 March 1990).

de Tailly, Martin. *Bruxella, Nobilissima Brabantiae Civitas.* Brussels: Bibliothèque Royale, Estampes, 1640(a), 1748(b).

Dear, M. S., and A. J. Scott, eds. *Urbanization and Urban Planning in Capitalist Society.* London: Methuen, 1981.

"Déclaration de Bruxelles." *Archives d'Architecture Moderne* 15 (1978). In *La reconstruction de Bruxelles: Recueil de projets publiés dans la Revue des Archives d'Architecture Moderne de 1977 à 1982.* Brussels: Editions des Archives d'Architecture Moderne, 1982.

Decroly, Jean-Michel, and Jean Pierre Grimmeau. "La démographie à l'échelle locale: Une géographie de la population de la Belgique dans les années 80." *Courrier Hebdomadaire du CRISP* [Centre de recherche et d'information socio-politique], nos. 1308–9 (1991).

Delattre, Bernard. "Quartier nord mode d'emploi." *La Libre Belgique,* Supplement hebdomadaire 17 (21 February 1991): 3.

Demey, Thierry. *Bruxelles: Chronique d'une capital en chantier; Du voûtement de la Senne à la jonction Nord-Midi.* Brussels: Paul Legrain/Edition CFC, 1990.

Denecke, Dieterich. "Research in German Urban Historical Geography." In *Urban Historical Geography: Recent Progress in Britain and Germany,* edited by Dieterich Denecke and Gareth Shaw. Cambridge: Cambridge University Press, 1988.

Denon, Vivant. *Discours sur les monuments d'antiquité d'Italie.* Paris: 8 vendémiaire an XII.

Des Marez, G. *Guide illustré de Bruxelles: Monuments civils et réligieux.* Brussels: Touring Club Royal de Belgique, 1979.

Despry, D. "La génèse d'une ville." In *Bruxelles: Croissance d'une capitale,* edited by J. Stengers. Antwerp, 1979.

Documents parlementaires. Sénat (1988–89), 514/1, p. 88; and 514/2, p. 118.

Documents parlementaires. Sénat (1979–80), 434/1, p. 1.

Doehaerd, Renée, Wim Blockmans, Hugo Soly, Els Witte, and Jan Craeybeckx. *Histoire de Flandre: Des origines à nos jours.* Brussels: La Renaissance du Livre, 1983.

Dogan, Matei, and John D. Kasarda, eds. *The Metropolis Era,* vol. 1, *A World of Giant Cities;* and vol. 2, *Mega-Cities.* Beverly Hills: Sage Publications, 1988.

Dorsett, L. W. *The Challenge of the City, 1860–1910.* Lexington: D. C. Heath, 1968.

Drumaux, Anne, Cecilia Maes, and Françoise Thys-Clément. "Bruxelles, les facteurs de l'équilibre budgétaire." *Courrier Hebdomadaire du CRISP* [Centre de recherche et d'information sociopolitique], nos. 1310–11 (1991).

Dupuis, Louis-André. *Plan topographique de la ville de Bruxelles et des ses environs.* Brussels: Bibliothèque Royale, Section des cartes et des plans IV, 6, 1777.

Eels, R., and C. Walton, eds. *Man in the City of the Future: A Symposium of Urban Philosophers.* New York: Arkville Press, 1968.

Eldredge, H. W., ed. *World Capitals: Toward Guided Urbanization.* New York: Anchor Press–Doubleday, 1975.

Eloy, Marc. *Influence de la législation sur les façades bruxelloises.* Brussels: Commission Française de la Culture de l'agglomeration de Bruxelles, 1985.

Elster, Jon. "Some Unresolved Problems in the Theory of Rational Behavior." *Acta Sociologica* 36 (1993): 179–90.

———. *Nuts and Bolts for the Social Sciences.* Cambridge: Cambridge University Press, 1989. Reprinted 1993.

———. *Making Sense of Marx.* Cambridge: Cambridge University Press, 1985.

———. *Ulysses and the Sirens.* Cambridge: Cambridge University Press, 1979.

European Cultural Foundation, ed. *Citizen and City in the Year 2000.* Rotterdam: Kluwer, Deventer, 1971.

European Parliament Building Service Statistics. Brussels: EP document no. 131.758/4.

The European Public Affairs Directory 1992. Brussels: Landmarks, S.A., 1991.

Evenson, Norma. "Paris, 1890–1940." In *Metropolis 1890–1940*, edited by Anthony Sutcliff. London: Mansel Publishers for Alexandrine Press, 1984.

———. *Paris: A Century of Change, 1978–1978*. New Haven: Yale University Press, 1979.

Fainstein, Susan S., et al. *Restructuring the City: The Political Economy of Urban Redevelopment*. New York: Longman, 1983.

Fenwick, T. [Jones Lang Wootton.] Handwritten note on completed survey. In author's possession. 1992.

Fishman, R. *Urban Utopias in the Twentieth Century: Ebenezer Howard, Frank Lloyd Wright, and Le Corbusier*. Cambridge: MIT Press, 1977.

Fitzmaurice, John. *The Politics of Belgium: Crisis and Compromise in a Plural Society*. London: C. Hurts & Co., 1988.

Fondation Roi Baudoin. *Approache cartographique des quartiers défavorisés de l'agglomeration bruxelloise: réagir à la dègradation socio-economique et urbanistique*. Brussels: La Fondation, 1984.

Forsyth, M. *Unions of States: The Theory and Practice of Confederation*. Leicester: Leicester University Press, 1981.

423 et 1 projets pour la Tête Défense: Le concours international d'architecture de 1983. Paris: Electa Moniteur, 1989.

Frank, Max. "La réforme de l'impôt sur le revenue: Problèmes d'équité." *Courrier Hebdomadaire du CRISP* [Centre de recherche et d'information socio-politique], no. 1236 (1989).

Freeman, Mike. "Commercial Building Development: The Agents of Change." In *The Built Form of Western Cities: Essays for M. R. G. Conzen on the Occasion of His Eightieth Birthday*, edited by T. R. Slater. Leicester: Leicester University Press, 1990.

Gauldie, E. *Cruel Habitations: A History of Working-Class Housing 1780–1918*. London: George Allen and Unwin, 1974.

Geddes, P. "The Twofold Aspect of the Industrial Age: Palaeotechnic and Neotechnic." *Town Planning Review* 31 (1912): 176–87.

Générale de Banque–Coopérative Ouvrière Belge. *Brochure of the Centre Internationale du Congrés*. Brussels: Générale de Banque–Coopérative Ouvrière Belge, 1987.

Gérard, Jo. *Télé-Bruxelles raconte Bruxelles*. Braine l'Alleud: J.-M. Collet, 1991.

Gibson, A. *People Power: Community and Work Groups in Action*. Harmondsworth: Penguin, 1979.

Goist, P. D. "Lewis Mumford and 'Anti-Urbanism.'" *Journal of the American Institute of Planners* 35 (1969): 340–47.

Goldstein, S., and D. F. Sly, eds. *Patterns of Urbanization: Comparative Country Studies*. Dolhain, Belgium: Ordina, 1977.

Gottlieb, M. "Battery Project Reflects Changing City Priorities." *New York Times,* 18 October 1985.

Green, C. M. *Washington: Capital City, 1879–1950.* Princeton: Princeton University Press, 1963.

Grimmeau, J.-P. "Caractéristiques fondamentales de l'éspace bruxellois." *Revue belge de géographie* 4 (1985).

Grunwald, Joseph, and Kenneth Flamm. *The Global factory: Foreign Assembly in International Trade.* Washington, DC: Brookings Institution, 1985.

Guicciardini, Lodovico. Engraved map of Brussels, ca. 1581–82 Knokke: Mappamundi Collection.

———. Woodcut map of Brussels, ca. 1567. Brussels: Bibliotheque Royale, Section des Livres précieux VB 10056 B.

Guide de l'architecture des années 25 à Bruxelles. Brussels: Archives d'Architecture Moderne, 1988.

Guillerme, Jacques. "Monument/Monumentalité: De la plénitude des symboles à la fascination du vide." In *L'illusion monumentale: Paris 1872–1936,* edited by Marilu Cantelli. Liège: Mardaga, 1991.

Habermas, J. "Modernity: An Incomplete Project," in *The Anti-Aesthetic,* edited by H. Foster. Port Townsend, WA: Bay Press, 1983.

Hague, C. *The Development of Planning Thought: A Critical Perspective.* London: Hutchinson, 1984.

Hall, P., and D. Hay. *Cities of Tomorrow: An Intellectual History of Urban Planning and Design in the Twentieth Century.* London: Basil Blackwell, 1988.

———. *The World Cities.* 3d ed. London: Weinfeld and Nicholson, 1984.

———. *Growth Centres in the European Urban System.* Berkeley: University of California Press, 1980.

———. *Great Planning Disasters.* London: Weinfeld and Nicholson, 1980.

Hammarström, I., and T. Hall, eds. *Growth and Transformation of the Modern City: The Stockholm Conference 1978.* Stockholm: Swedish Council for Building Research, 1979.

Handlin, Oscar, and John Burchard, eds. *The Historian and the City.* Cambridge, MA: MIT Press, 1963.

Harvey, David. "Between Space and Time: Reflections on the Geographical Imagination." *Annals of the Association of American Geographers* 80 (1990): 418–34.

——— . "*Urban Fortunes: The Political Economy of Place;* Review." *Environment and Planning D: Society and Space* 8 (1990): 495–96.

——— . *The Condition of Postmodernity.* Oxford: Blackwell, 1989.

————. "Flexible Accumulation through Urbanization: Reflections on 'Post-Modernism' in the American City." *Antipode* 19, 3 (1987): 260–61.

————. *Consciousness and the Urban Experience: Studies in the History and Theory of Capitalist Urbanization.* Baltimore: Johns Hopkins University Press, 1985.

————. *The Urbanization of Capital: Studies in the History and Theory of Capitalist Urbanization.* Baltimore: Johns Hopkins University Press, 1985.

————. *Social Justice and the City.* London: Edward Arnold, 1973.

Hasquin, Hervé, ed. *Communes de Belgique: Dictionnaire d'histoire et de géographie administrative; Wallonie-Bruxelles.* Brussels: La renaissance du livre pour le Crédit Communal de Belgique, 1983.

Hay, D., and P. Hall. *Urban Regionalization in Belgium 1970: Working Paper 3.* Reading: Reading University, Geography Department, 1976.

Healey & Baker. *Belgium.* Brussels: Healey & Baker, 1990.

Heinritz, Günter, and Elisabeth Lichtenberger, eds. *The Take-Off of Suburbia and the Crisis of the Central City: Proceedings of the International Symposium in Munich and Vienna 1984.* Stuttgart: Franz Steiner Verlag Wiesbaden Gmbh, 1986.

Institut Géographique National. Air photograph of the quartiers Léopold and Nord-Est (detail), sheet nos. 31–32/1317, 1991.

————. Air photograph of quartier Léopold and quartier Nord-Est (detail), sheet nos. 30–31/1351, 1988.

————. Air photograph of quartier Léopold and quartier Nord-Est (detail), sheet no. 31/1314, 1978.

————. Air photograph of quartier Léopold and quartier Nord-Est (detail), sheet no. 31/1313, 1969.

————. Air photograph of quartier Léopold and quartier Nord-Est (detail), sheet no. 31/201, 1956.

————. Air photograph of quartier Léopold and quartier Nord-Est (detail), sheet no. 31/467, 1950.

Institut National du Logement. *Enquête sur la démolition des logements insalubres et le relogement de laurs occupants.* Brussels: Institut National du Logement, 1971.

Institut de Sociologie de l'Université Libre de Bruxelles; Centre Européen pour Bruxelles; Centre d'études et de recherches urbaines; Présidence de l'Executif de la Communauté Française. *Quels devenirs pour Bruxelles et sa région? Actes du colloque, 5, 6, 9 décembre 1983.* Brussels: Editions de l'Université de Bruxelles, 1984.

Jacobs, J. *The Death and Life of Great American Cities.* London: Jonathan Cape, 1962.

Jacqmain, C. "Les operations de la société anonyme du quartier Notre-Dame-aux-Neiges (1984–1888)." 2 vols. Thesis, Université Libre de Bruxelles, 1976.

Janssens, Luc, and Lisette Danckaert. "La grande propriété immobilière et son evolution." In *La région de Bruxelles: Des villages d'autrefois à la ville d'aujourd'hui,* edited by Arlette Smolar-Meynart and Jean Stengers, pp. 196–211. Brussels: Credit Communal, Collection Histoire Serie no. 16, 1989.

Jones Lang Wootton International. "Worldwide Office Occupancy Costs." *JLW Research* 1 (September 1989).

Jouret, Bernard. *Définition spatiale du phénomène urbain Bruxellois.* Brussels: Editions de l'Université de Bruxelles, 1972.

Kesteloot, Christian. "Three Levels of Socio-Spatial Polarization in Brussels." *Built Environment* 20, no. 3 (1994): 204–17.

King, Anthony D. *Global Cities: Post-Imperialism and the Internationalization of London.* London: Routledge, 1990.

Knox, Paul. "Capital, Material Culture and Socio-Spatial Differentiation." In *The Restless Urban Landscape,* edited by Paul Knox. Englewood Cliffs, NJ: Prentice Hall, 1993.

Koter, Marek. "The Morphological Evolution of a Nineteenth-Century City Centre: Lódz, Poland (1825–1973)." In *The Built Form of Western Cities: Essays for M. R. G. Conzen on the Occasion of His Eightieth Birthday,* edited by T. R. Slater. Leicester: Leicester University Press, 1990.

La Grande Arche c'est tout Paris [advertising brochure]. Paris: Saco & Co., 1990.

La réforme de l'Etat. Brussels: Centre d'études pour la réforme de l'état, 1937.

La tour ferrée: Projets dans la ville; projets réalisés à l'institut supérieur d'architecture, La Cambre, Bruxelles de 1975 à 1978. Brussels: Archives d'Architecture Moderne, 1978.

Lake, Robert. "*Urban Fortunes: The Political Economy of Place;* A Commentary." *Urban Gerography* 11, no. 2 (1990).

Larkham, Peter J., and Andrew N. Jones. *A Glossary of Urban Form.* Leicester, UK: Historical Geography Research Series, no. 26, 1991.

Lavedan, P. *Histoire de l'urbanisme à Paris.* Paris: Hachette, 1975.

———. *Géographie des villes.* Paris: Gallimard, 1959.

———. *Histoire d'urbanisme: Epoque contemporaine.* Paris: Henri Laurens, 1952.

"Le logement ouvrier dans l'impasse?" *Les cahiers de la fonderie: Revue d'histoire social et industrielle de la région bruxelloise* 6 (June1989).

Lees, A. *Cities Perceived: Urban Society in European and American Thought, 1820–1940.* Manchester: Manchester University Press, 1966.

Lefèbvre, H. *Espace et politique: Le droit à la ville II.* Paris: Editions Anthropos, 1972.

———. *Le droit à la ville.* Paris: Editions Anthropos, 1968.

———. "Space: Social Product and Use Value." In *Critical Sociology: European Perspectives,* edited by J. W. Freiberg, pp. 285–95. New York: Irvington Publishers, 1979.

Léfèvre, P. "Le problème de la paroisse primitive de Bruxelles." *Annales de la société royale d'archeologie de Bruxelles* 38 (1934): 106–16.

Lemaître, H. *Les gouvernements belges de 1968 à 1980: Processus de crise.* Brussels: Chauveheid, 1982.

Léopold II. "Embellissement de Bruxelles (1): Discours du Roi à l'inauguration de l'Exposition, 7 juin 1888." Brussels, Archives du Palais royal, Cabinet de Léopold II.

Leven, Charles, ed. *The Mature Metropolis.* Lexington, MA: Lexington Books, 1978.

Lewis, D. N., ed. *The Growth of Cities* (Architects' Year Book, XIII). London: Elek Books, 1971.

Lichtenberger, Elisabeth. *Vienna: Bridge Between Cultures.* London: Belhaven Press, 1993.

———. "Political Systems and Urban Development." *Geografische Tijdschrift* 25, no. 5 (1991): 421–26.

———. "Perspectives in Urban Geography." *Recherches de Géographie urbaine, Hommage au Professeur Sporck,* vol. 1. Liège: Press Universitaire de Liège, 1987.

———. "The Changing Nature of European Urbanism." In *Urbanization and Counter-Urbanization* (Urban Affairs Annual Reviews vol. 11), edited by Brian B. Berry, pp. 81–107. Beverly Hills: Sage, 1976.

———. "The Nature of European Urbanism." *Geoforum* 4 (1970): 45–62.

Linteau, Paul André. *Maisonneuve: Comment des promoteurs fabriquent une ville, 1883–1918.* Montréal: Boréal Express, 1981.

Logan, John. "How to Study the City: Arguments for a New Approach." *Urban Geography* 11, no. 2 (1990): 201.

Logan, John, and Harvey L. Molotch. *Urban Fortunes: The Political Economy of Place.* Berkeley: University of California Press, 1987.

Loumaye, Serge. "Les nouvelles institutions bruxelloises." *Courrier Hebdomadaire du CRISP* [Centre de recherche et d'information sociopolitique], nos. 1232–33 (1989).

Löwy, E. "Die Anfäge des Triumphbogens." *Jahrbuch der kunsthistorischen Sammlungen in Wien*, N.F. vol. 2 (1928).

Loze, Pierre, ed. *Guide de Bruxelles XIXème & art nouveau.* Brussels: Eiffel Editions, 1990.

Lynch, Kevin. *Good City Form.* Cambridge, MA: MIT Press, 1981.

Mabille, Xavier, Evelyn Lentzen, and Pierre Blaise. "Les élections du 24 novembre 1991." *Courrier Hebdomadaire du CRISP* [Centre de recherche et d'information socio-politique], nos. 1335–36 (1991).

Mardaga, Pierre, ed. *Le patrimoine monumental de la Belgique: Bruxelles*, vol. 1, book A, *Pentagone A-D.* Liège: Solédi pour les Communautés Française et Flamande, 1989.

Martens, M. "Les survivances domaniales du 'castrum' carolignien de Bruxelles à la fin de Moyen Age." *Le Moyen Age* 69 (1963): 641–55.

Masser, I. "An Emerging World City." *Town and Country Planning* 49 (1980): 301–3.

Massey, D. "Enterprise Zones: A Political Issue." *International Journal of Urban and Regional Research* 6 (1982): 429–34.

May, E. "Cities of the Future." *Survey* 38 (1961): 179–85.

McIntyre, Michael. "The Chambonnais Paradox." *Polity* (forthcoming).

McRae, K. *Conflict and Compromise in Multilingual Societies: Belgium.* Waterloo, Ontario: Wilfried Laurier University Press, 1986.

Méan, A. "La casse-tête des matières personalisable à Bruxelles." *Belgique: la révision constitutionnelle et la régionalisation; Problèmes économiques, politiques, et sociaux.* Brussels: La Documentation Française, 1981.

Mens en Ruimte. *Brussel, de internationale uitdaging: De direct sociaaleconomische impact van de internationale basisinstellingen in Brussels.* Brussels: Mens en Ruimte v.z.w., 1990.

Mills, Edwin S., and Bruce W. Hamilton. *Urban Economics.* 3d ed. Glenview, IL: Scott, Foresman and Co., 1984.

Ministerie van de Vlaamse Gemeenschap, Bestuur voor monumenten en landscappen; Ministère de la Communauté Française, Administration du Patrimoine Culturel, *Le patrimoine monumental de la Belgique: Bruxelles*, vol. 1, book 1, Pentagone A-D. Brussels: Pierre Mardaga.

Mitchell, B. R. *European Historical Statistiques, 1750–1970.* London: Macmillan, 1975.

Moke, Henri. *La Belgique monumentale, historique et pittoresque*, book 1. Brussels, 1844.

Morizet, A. *Du vieux Paris au Paris moderne: Haussmann et ses prédécesseurs.* Paris: Hachette, 1932.

Mumford, Lewis. *The City in History: Its Origins, Its Transformations, and Its Prospects*. New York: Harcourt, Brace, 1961.

Murphy, Alexander B. "Ethno-nationalism and the Social Construction of Space." Paper, Toronto Meeting of the *Association of American Geographers*, 1990.

———. *The Regional Dynamics of Language Differentiation in Belgium: A Study in Cultural-Political Geography*. Chicago: University of Chicago Geography Research Paper no. 227, 1988.

Nilsson, Martini P. *Opuscula Selecta*. 2 vols. Lund: Skrifter Utgivna av Svenka Institutet i Athen. CWK Gleerup, 1952.

———. "The Origin of the Triumphal Arch," and "The Triumphal Arch and Town Planning." In his *Opuscula Selecta*. Lund: Skrifter Utgivna av Svenka Institutet i Athen. CWK Gleerup, 1952.

———. "Les bases votives à double colonne et l'arc de triomphe." *Bulletin de correspondance hellenique* 49 (1925): 143.

Noak, F. "Triumph und Triumpbogen." *Vorträge der Bibliothek Warburg* 26 (1925).

Office des Propriétaires Immobilier S.A. *Le Marché Bruxellois*. Brussels: Etude de la société OP, 1992.

Olympia & York. *Canary Wharf: The Untold Story*. London: Olympia & York, 1990.

Pelli, Cesar. "The Mega Building on Context." *Architectural Design* 58, 11/12 (1989): 40–44.

Peterson, Paul E. *City Limits*. Chicago: University of Chicago Press, 1981.

Pinkney, David H. *Napoleon III and the Rebuilding of Paris*. Princeton, NJ: Princeton University Press, 1972.

Poete, M. *Une vie de cité: Paris de sa naissance à nos jours*. 3 vols. Paris: Auguste Picard, 1931.

Portoghesi, P. *Postmodern: The Architecture of the Postindustrial Society*. New York: Rizzoli, 1982.

Posner, Kenneth. "An Amenity Analysis of Office Rents in the Chicago Central Business District." *Journal of the American Real Estate and Urban Economics Association*. Forthcoming.

Pounds, N. J. G. *An Historical Geography of Europe*. Cambridge, UK: Cambridge University Press, 1990.

Puttemans, Pierre. "L'architecture de la ville." *Questions* 10 (1991).

———. *L'architecture moderne en Belgique*. Brussels: Vokaer, 1974.

R., E. "Grands projets immobiliers." *La Denière Heure* (6 May 1991).

Ragon, Michel. *Histoire de l'architecture et de l'urbanisme modernes*. Paris: Casterman, 1986.

Ranieri, Liane. *Léopold II: Urbaniste*. Brussels: Hayez, 1973.

Rapoport, Amos. *The Meaning of the Built Environment: A Nonverbal Communication Approach.* Tucson: University of Arizona Press, 1990.

"Rapport Spécial de la Cour des Comptes relatif à la politique immobilière des institutions des Communautés Européennes." *Journal Officiel des Communautés Européennes* 221, no. 2 (3 September 1979): 11.

Ray, Richard. [Manager, Investment Department of Richard Ellis.] Letter, Brussels, 2 October 1992.

Region de Bruxelles-Capitale. *Stratégie pour une géstion regionale de la fonction administrative.* Administration de l'urbanisme et de l'aménagement du territoire. Document de synthèse. Brussels: BRAT, 1991.

Renoy, Georges. *Bruxelles vecu: Quartier royal.* Brussels: Rossel, 1980.

"Répertoire Télécom." Internal document of the Commission des C.E., Brussels, February 1985.

Rimmer, Peter. "Japanese Construction Contractors and the Australian States: Another Round of Interstate Rivalry." *International Journal of Urban and Regional Research* 12, no. 3 (1988): 404–24.

Robinson, C. M. *The Improvement of Towns and Cities; Or, The Practical Basis of Civic Aesthetics.* New York: G. P. Putnum's Sons, 1901.

Robson, D. "Canary Wharf." *Planner* (October 12, 1990): 8–10.

Robson, W. A., and D. E. Regan, eds. *Great Cities of the World.* 3d ed. 2 vols. London: George Allen and Unwin, 1972.

Rodger, R. G. "Sources and Methods of Urban Studies: The Contribution of Building Records." *Area* 13 (1981): 315–21.

Rosenthal, D. B., ed. *Urban Revitalization* (Urban Affairs Annual Reviews, no. 18). Beverly Hills: Sage, 1980.

Sandercock, Leonie. *Property, Politics, and Urban Planning: The History of Australian City Planning, 1890–1990.* New Brunswick, NJ: Transaction Publishers, 1990.

Schelling, Thomas C. *The Strategy of Conflict.* Cambridge, MA: Harvard University Press, 1960.

Schneider, Rudolph. [European Commission, Directorate General XXIII: Enterprise Policy, Cooperatives, and Tourism.] Interview, Brussels, Belgium, 13 February 1992.

Schoonbrodt, Réné. [ARAU.] Interview, Brussels, Belgium, 10 September 1991.

———. *Unifier les institutions qui gouvernent Bruxelles pour que la ville ait tout pouvoir sur elle-même.* Brussels: ARAU, 1988.

———. *Essai sur la destruction des villes et des campagnes.* Brussels: Pierre Mardaga, Série Architecture et Recherche, 1987.

————. *Sociologie de l'habitat social: Comportement des habitants et architecture des cités.* Brussels: Éditions des Archives d'Architecture Moderne, 1979.

Schultz, S. K., and C. McShane. "To Engineer the Metropolis: Sewers, Sanitation, and City Planning in Late-Nineteenth Century America." *Journal of American History* 65 (1978): 389–411.

Schuyler, David. *The New Urban Landscape: The Redefinition of City Form in Nineteenth-Century America.* Baltimore: Johns Hopkins University Press, 1986.

Scott, A. J., and S. T. Roweis. "Urban Planning in Theory and Practice: An Appraisal." *Environment and Planning A* 9 (1977): 1097–1119.

Secretariat d'état à la région Bruxelloise. Administration de l'urbanisme et de l'aménagement du territoire. *Plan de secteur de l'agglomeration Bruxelloise: Situation existante de droit et de fait.* Brussels, 1979.

Sellier, H., and A. Bruggeman. *Le problème de logement: Son influence sur les conditions de l'habitations et l'aménagement des villes.* Paris: Presses Universitaires de France, 1927.

Semal, Harold. [Ascona & Co.] Interview, 3 March 1990, Brussels, Belgium.

Sint-Lukasarchief; Commission néerlandaise de la culture de l'agglomeration de Bruxelles; Crédit Communal de Belgique, eds. *Bruxelles, construire et reconstruire: Architecture et aménagement urbain 1780–1914.* Brussels: Crédit Communal de Belgique, 1979.

Sitte, Camillo. *L'art de batir les villes: L'urbanisme selon ses fondements artistiques.* Paris: Livre et Communication, 1990.

Slater, T. R. "Starting Again: Recollections of an Urban Morphologist." In *The Built Form of Western Cities: Essays for M. R. G. Conzen on the Occasion of His Eightieth Birthday*, edited by T. R. Slater. Leicester: Leicester University Press, 1990.

————. "Urban Morphology in 1990: Developments in International Co-operation." In *The Built Form of Western Cities: Essays for M. R. G. Conzen on the Occasion of His Eightieth Birthday*, edited by T. R. Slater. Leicester: Leicester University Press, 1990.

Smith, Michael Peter. *City, State, and Market: The Political Economy of Urban Society.* New York: Basil Blackwell, 1988.

Smolar-Meynart, A., and J. Stengers. *La région de Bruxelles: Des villages d'autrefois à la ville d'aujourd'hui*, Collection Histoire Series, no. 16. Brussels: Crédit Communal de Belgique, 1989.

Société Centrale d'Architecture de Belgique, ed. *Bruxelles: Reflexion prospective sur l'agglomeration*, vols. 1-2. Brussels: Ministère de la Communauté Française, Administration du patrimoine culturel, 1980.

Soja, Edward W. *Postmodern Geographies: The Reassertion of Space in Critical Social Theory.* London: Verso, 1989.

Spinelli, A. "European Union in the Resistance." *Government and Opposition* 2 (1966–67).

Staes, Paul. *Europe at Stake in the Speculation Game: A Political and Ecological Analysis of a Billion-Franc Project; The European Parliament Plays Pontius Pilate.* Brussels: European Parliament document no. 95374, September 1990.

Stauven, Francis. "L'ideologie du modernisme belge après l'Art Nouveau." *L'architectura in Belgio: 1920–1940,* 32, no. 2, 1979.

Stave, B. M., ed. *Modern Industrial Cities: History, Policy, and Survival.* Beverly Hills: Sage, 1981.

Stynen, Herman. *Urbanisme et Société: Louis Van der Swaelmen (1883–1929) animateur du mouvement moderne en Belgique.* Liège: Pierre Mardaga, 1979.

Sutcliffe, A., and A. Sutcliffe. *The Autumn of Central Paris: The Defeat of Town Planning 1850–1970.* London: Edward Arnold, 1970.

Suys, T. *Plan du Palais de Justice à construire à Bruxelles.* Brussels, 1838.

Tabb, W. K., and L. Sawers, eds. *Marxism and the Metropolis: New Perspectives in Urban Political Economy,* 2d ed. New York: Oxford University Press, 1984.

Taylor, P. *The Limits of European Integration.* London: Croom Helm, 1983.

Thiry, Jean-Pierre, ed. *L'Europe à Bruxelles.* Brussels: Centre d'études et de recherches Urbaines–ERU a.s.b.l.

Timmerman, Georges. *Main basse sur Bruxelles: Argent, pouvoir, et béton.* Brussels: Editions EPO, 1991.

Tolley, George, and William Shear, eds. *Housing Dynamics and Neighborhood Change.* Chicago: Blackstone Books, Studies in Urban and Resource Economics, 1986.

Tolley, George, and Shou-yi Hao. "Urban Land Use and Land Prices in Market Economies." 1992. Typescript.

Treaties Establishing the European Communities: Treaties Amending These Treaties; Single European Act. Luxembourg: Office for Official Publications of the European Communities, 1987.

Tugendhat, C. *Making Sense of Europe.* London: Viking Penguin, 1986.

Tugendhat, C., and W. Van Waelvelde. "Bilans de Main-d'Oeuvre et Mouvement des Migrants Alternants." *Bulletin de Statistique* 6, no. 3 (1974).

Van der Haegen, H., and M. Pattyn. "Les Régions Urbaines Belges." *Bulletin de Statistique* 3 (1979): 235–49.

van Deventer, Jacob. Manuscript map of Brussels and its environs, ca. 1550. Brussels: Bibliotheque Royale, Section des Manuscrits 22090.

Van Mierlo, C. *Projet d'ensemble pour l'amelioration de la voirie et la transformation de divers quartiers.* Brussels, 1885.

Van Parys, Germaine. *Pas perdus dans Bruxelles: Photographies du début du siècle.* Brussels: Monique Adam et Agence de Press Van Parys, 1979.

Vandenbossche, A. M. *Les bureaux à Bruxelles.* Brussels: Ministère de la Région Bruxelloise, 1988.

Vanhamme, Marcel. *Bruxelles: De bourg rural à cité mondiale.* Brussels: Mercurius, 1978.

Vantroyen, Jean-Claude. "La façade de Gustave Saintenoy, un détail?" *Le Soir* (17 May 1990).

Verwest, Auguste. *Nouveau plan de Bruxelles industriel avec ses environs.* Brussels: Plans industriels de Belgique, Khiat & Co., 1910.

Vigneron-Zwetkoff, C. "L'administration des grandes agglomérations belges." *Administration Public* (1983): 142.

Ville de Bruxelles. *Règlement sur les bâtisses.* Brussels, 1981.

Vincent, Anne. "Les investissements nordiques en Belgique." *Courrier Hébdomadaire du CRISP* [Centre de recherche et d'information sociopolitique], nos. 1246–47 (1989).

W., L. "Rebâtir le carrefour de l'Europe: L'ARAU soutient trois projets de reconstruction de logements sur le site, mais veut en améliorer l'architecture." *La Lanterne* (3 May 1991).

Wallace, H., W. Wallace, and C. Webb, eds. *Policy Making in the European Community.* London: John Wiley, 1983.

Wheatley, Paul. *Nāgara and Commandery: Origins of the Southeast Asian Urban Traditions.* Chicago: University of Chicago Geography Research Paper nos. 207–8, 1983.

———. *City as Symbol: An Inaugural Lecture Delivered at University College, London, 20 November 1967.* London: H. K. Lewis & Co. for the University College, 1969.

Whitehand, J. W. R. *The Making of the Urban Landscape.* Oxford: Blackwell Publishers, 1993.

———. *The Changing Face of Cities: A Study of Development Cycles and Urban Form.* Oxford: Basil Blackwell, 1987.

———. *Rebuilding Town Centres: Developers, Architects, and Styles.* Birmingham, UK: Department of Geography, University of Birmingham Occasional Publication no. 19, 1984.

———. "Background to the Urban Morphogenetic Tradition." In *The Urban Landscape: Historical Development and Management: Papers by*

M. R. G. Conzen. London: Academic Press, Institute of British Geographers, Special Publication no. 13, 1981.

Wigny, P. *La troisième révision de la Constitution.* Brussels: Bruylant, 1972.

Williams, Charles R. "Regional Management Overseas." *Harvard Business Review* (January-February 1967): 87–91.

Willis, Virginia. *Britons in Brussels: Officials in the European Commission and Council Secretariat.* London: European Centre for Political Studies and the Royal Institute of Public Administration, 1982.

Witte, Els, and Hugo Baetens Beardsmore, eds. *The Interdisciplinary Study of Urban Bilingualism in Brussels.* Philadelphia, PA: Multilingual Matters, 1987.

Zoller, Henry G. *Localisation résidentielle: Décision des ménages et développement suburbain.* Brussels: Les Editions Vie Ouvrière, 1972.

Index

The University of Chicago
GEOGRAPHY RESEARCH PAPERS

Titles in Print

127. GOHEEN, PETER G. *Victorian Toronto, 1850 to 1900: Pattern and Process of Growth.* 1970. xiii + 278 pp.

132. MOLINE, NORMAN T. *Mobility and the Small Town, 1900–1930.* 1971. ix + 169 pp.

136. BUTZER, KARL W. *Recent History of an Ethiopian Delta: The Omo River and the Level of Lake Rudolf.* 1971. xvi + 184 pp.

152. MIKESELL, MARVIN W., ed. *Geographers Abroad: Essays on the Problems and Prospects of Research in Foreign Areas.* 1973. ix + 296 pp.

181. GOODWIN, GARY C. *Cherokees in Transition: A Study of Changing Culture and Environment Prior to 1775.* 1977. ix + 207 pp.

186. BUTZER, KARL W., ed. *Dimensions of Human Geography: Essays on Some Familiar and Neglected Themes.* 1978. vii + 190 pp.

194. HARRIS, CHAUNCY D. *Annotated World List of Selected Current Geographical Serials, Fourth Edition. 1980.* 1980. iv + 165 pp.

206. HARRIS, CHAUNCY D. *Bibliography of Geography. Part II: Regional. Volume 1. The United States of America.* 1984. viii + 178 pp.

207–208. WHEATLEY, PAUL. *Nagara and Commandery: Origins of the Southeast Asian Urban Traditions.* 1983. xv + 472 pp.

209. SAARINEN, THOMAS F.; DAVID SEAMON; and JAMES L. SELL, eds. *Environmental Perception and Behavior: An Inventory and Prospect.* 1984. x + 263 pp.

210. WESCOAT, JAMES L., JR. *Integrated Water Development: Water Use and Conservation Practice in Western Colorado.* 1984. xi + 239 pp.

213. EDMONDS, RICHARD LOUIS. *Northern Frontiers of Qing China and Tokugawa Japan: A Comparative Study of Frontier Policy.* 1985. xi + 209 pp.

216. OBERMEYER, NANCY J. *Bureaucrats, Clients, and Geography: The Bailly Nuclear Power Plant Battle in Northern Indiana.* 1989. x + 135 pp.

217–218. CONZEN, MICHAEL P., ed. *World Patterns of Modern Urban Change: Essays in Honor of Chauncy D. Harris.* 1986. x + 479 pp.

222. DORN, MARILYN APRIL. *The Administrative Partitioning of Costa Rica: Politics and Planners in the 1970s.* 1989. xi + 126 pp.

223. ASTROTH, JOSEPH H., JR. *Understanding Peasant Agriculture: An Integrated Land-Use Model for the Punjab.* 1990. xiii + 173 pp.

224. PLATT, RUTHERFORD H.; SHEILA G. PELCZARSKI; and BARBARA K. BURBANK, eds. *Cities on the Beach: Management Issues of Developed Coastal Barriers.* 1987. vii + 324 pp.

225. LATZ, GIL. *Agricultural Development in Japan: The Land Improvement District in Concept and Practice.* 1989. viii + 135 pp.

226. GRITZNER, JEFFREY A. *The West African Sahel: Human Agency and Environmental Change.* 1988. xii + 170 pp.

227. MURPHY, ALEXANDER B. *The Regional Dynamics of Language Differentiation in Belgium: A Study in Cultural-Political Geography.* 1988. xiii + 249 pp.

228–229. BISHOP, BARRY C. *Karnali under Stress: Livelihood Strategies and Seasonal Rhythms in a Changing Nepal Himalaya.* 1990. xviii + 460 pp.

230. MUELLER-WILLE, CHRISTOPHER. *Natural Landscape Amenities and Suburban Growth: Metropolitan Chicago, 1970–1980.* 1990. xi + 153 pp.

231. WILKINSON, M. JUSTIN. *Paleoenvironments in the Namib Desert: The Lower Tumas Basin in the Late Cenozoic.* 1990. xv + 196 pp.

232. DUBOIS, RANDOM. *Soil Erosion in a Coastal River Basin: A Case Study from the Philippines.* 1990. xii + 138 pp.

233. PALM, RISA, AND MICHAEL E. HODGSON. *After a California Earthquake: Attitude and Behavior Change.* 1992. xii + 130 pp.

234. KUMMER, DAVID M. *Deforestation in the Postwar Philippines.* 1992. xviii + 179 pp.

235. CONZEN, MICHAEL P., THOMAS A. RUMNEY, AND GRAEME WYNN. *A Scholar's Guide to Geographical Writing on the American and Canadian Past.* 1993. xiii + 751 pp.

236. COHEN, SHAUL EPHRAIM. *The Politics of Planting: Israeli-Palestinian Competition for Control of Land in the Jerusalem Periphery.* 1993. xiv + 203 pp.

237. EMMETT, CHAD F. *Beyond the Basilica: Christians and Muslims in Nazareth.* 1994. xix + 303 pp.

238. PRICE, EDWARD T. *Dividing the Land: Early American Beginnings of Our Private Property Mosaic.* 1995. xviii + 410 pp.

239. PAPADOPOULOS, ALEX G. *Urban Regimes and Strategies: Building Europe's Central Executive District in Brussels.* 1996. xviii + 290 pp.